Boosting-Based Face Detection and Adaptation

Synthesis Lectures on Computer Vision

Editor

Gérard Medioni, *University of Southern California*
Sven Dickinson, *University of Toronto*

Synthesis Lectures on Computer Vision is edited by Gérard Medioni of the University of Southern California and Sven Dickinson of the University of Toronto. The series will publish 50- to 150 page publications on topics pertaining to computer vision and pattern recognition. The scope will largely follow the purview of premier computer science conferences, such as ICCV, CVPR, and ECCV. Potential topics include, but not are limited to: Applications and Case Studies for Computer Vision Color, Illumination, and Texture; Computational Photography and Video; Early and Biologically-inspired Vision; Face and Gesture Analysis; Illumination and Reflectance Modeling; Image-Based Modeling; Image and Video Retrieval; Medical Image Analysis; Motion and Tracking; Object Detection, Recognition, and Categorization; Segmentation and Grouping; Sensors; Shape-from-X; Stereo and Structure from Motion; Shape Representation and Matching; Statistical Methods and Learning; Performance Evaluation; and Video Analysis and Event Recognition.

Boosting-Based Face Detection and Adaptation
Cha Zhang and Zhengyou Zhang
2010

Image-Based Modeling of Plants and Trees
Sing Bing Kang, Long Quan
2009

Boosting-Based Face Detection and Adaptation

Cha Zhang and Zhengyou Zhang

www.morganclaypool.com

ISBN: 9781608451333   paperback
ISBN: 9781608451340   ebook

DOI 10.2200/S00300ED1V01Y201009COV002

A Publication in the Morgan & Claypool Publishers series
SYNTHESIS LECTURES ON COMPUTER VISION

Lecture #2
Series Editor: Gérard Medioni, *University of Southern California*
                Sven Dickinson, *University of Toronto*
Series ISSN
Synthesis Lectures on Computer Vision
Print 2153-1056   Electronic 2153-1064

Boosting-Based Face Detection and Adaptation

Cha Zhang and Zhengyou Zhang
Microsoft Research

SYNTHESIS LECTURES ON COMPUTER VISION #2

MORGAN & CLAYPOOL PUBLISHERS

ABSTRACT

Face detection, because of its vast array of applications, is one of the most active research areas in computer vision. In this book, we review various approaches to face detection developed in the past decade, with more emphasis on boosting-based learning algorithms. We then present a series of algorithms that are empowered by the statistical view of boosting and the concept of multiple instance learning.

We start by describing a boosting learning framework that is capable to handle billions of training examples. It differs from traditional bootstrapping schemes in that no intermediate thresholds need to be set during training, yet the total number of negative examples used for feature selection remains constant and focused (on the poor performing ones). A multiple instance pruning scheme is then adopted to set the intermediate thresholds after boosting learning. This algorithm generates detectors that are both fast and accurate.

We then present two multiple instance learning schemes for face detection, multiple instance learning boosting (MILBoost) and winner-take-all multiple category boosting (WTA-McBoost). MILBoost addresses the uncertainty in accurately pinpointing the location of the object being detected, while WTA-McBoost addresses the uncertainty in determining the most appropriate subcategory label for multiview object detection. Both schemes can resolve the ambiguity of the labeling process and reduce outliers during training, which leads to improved detector performances.

In many applications, a detector trained with generic data sets may not perform optimally in a new environment. We propose detection adaption, which is a promising solution for this problem. We present an adaptation scheme based on the Taylor expansion of the boosting learning objective function, and we propose to store the second order statistics of the generic training data for future adaptation. We show that with a small amount of labeled data in the new environment, the detector's performance can be greatly improved.

We also present two interesting applications where boosting learning was applied successfully. The first application is face verification for filtering and ranking image/video search results on celebrities. We present boosted multi-task learning (MTL), yet another boosting learning algorithm that extends MILBoost with a graphical model. Since the available number of training images for each celebrity may be limited, learning individual classifiers for each person may cause overfitting. MTL jointly learns classifiers for multiple people by sharing a few boosting classifiers in order to avoid overfitting. The second application addresses the need of speaker detection in conference rooms. The goal is to find who is speaking, given a microphone array and a panoramic video of the room. We show that by combining audio and visual features in a boosting framework, we can determine the speaker's position very accurately.

Finally, we offer our thoughts on future directions for face detection.

KEYWORDS

face detection, boosting, multiple instance learning, adaptation, multiple task learning, multimodal fusion

To Jing, Alex, and Eric
– C.Z.

To Ming-Yue, Rosaline, Laetitia, and Stephanie
– Z.Z.

Contents

Preface

With the rapid increase of computational powers and availability of modern sensing, analysis and rendering equipment and technologies, computers are becoming more and more intelligent. Many research projects and commercial products have demonstrated the capability for a computer to interact with human in a natural way by looking at people through cameras, listening to people through microphones, understanding these inputs, and reacting to people in a friendly manner.

One of the fundamental techniques that enables such natural human-computer interaction (HCI) is face detection. Face detection is the step stone to all facial analysis algorithms, including face alignment, face modeling, face relighting, face recognition, face verification/authentication, head pose tracking, facial expression tracking/recognition, gender/age recognition, and many many more. Only when computers can understand face well will they begin to truly understand people's thoughts and intentions.

It is a non-trivial task for a computer to detect faces in images or videos, and it has been one of the most studied topics in the computer vision literature. It was not until the seminal work by Viola and Jones (2001) that face detection became widely used in real world applications. Today, if you buy a digital camera, most likely, it comes with face detection software to help auto-focusing and auto-exposure. These face detectors works reasonably well, though challenges still remain due to the huge variations of face images in scale, location, lighting, head pose, etc.

The authors of this book have worked intensely on the face detection problem in the past 4-5 years to further improve the original Viola-Jones detector (Viola and Jones, 2001). This book is based on a number of our publications such as (Viola et al., 2005), (Zhang and Viola, 2007), (Zhang et al., 2008a), (Zhang et al., 2008b), (Zhang and Zhang, 2009), (Wang et al., 2009b), etc. It is our hope that by sharing some of the lessons we learned during our exploration, we will see even better algorithms developed to solve this fundamental computer vision problem.

We would like to take this opportunity to give our special thanks to Paul Viola, one of the designers of the original boosting-based face detector, with whom we developed some of the key algorithms presented in this book. We would also like to thank other co-authors of those papers, namely, John Platt, Raffay Hamid, Pei Yin, Yong Rui, Ross Cutler, Xinding Sun, Nelson Pinto, and Xiaogang Wang. We would also like to thank Rong Xiao, with whom we had many valuable discussions while developing some of the algorithms presented in this book.

Cha Zhang and Zhengyou Zhang
September 2010

CHAPTER 1

A Brief Survey of the Face Detection Literature

1.1 INTRODUCTION

Given an arbitrary image, the goal of face detection is to determine whether or not there are any faces in the image and, if present, return the image location and extent of each face (Yang et al., 2002). While this appears as a trivial task for human beings, it is a very challenging task for computers and it has been one of the most heavily studied research topics in the past few decades. The difficulty associated with face detection can be attributed to many variations in skin color, scale, location, orientation (in-plane rotation), pose (out-of-plane rotation), facial expression, lighting conditions, occlusions, etc., as seen in Fig. 1.1.

There have been hundreds of reported approaches to face detection. Early works (before year 2000) had been nicely surveyed in Yang et al. (2002) and Hjelmas and Low (2001). For instance, Yang et al. (2002) grouped the various methods into four categories: knowledge-based methods, feature invariant approaches, template matching methods, and appearance-based methods. Knowledge-based methods use pre-defined rules to determine a face based on human knowledge; feature invariant approaches aim to find face structure features that are robust to pose and lighting variations; template

Figure 1.1: Examples of face images. Note the huge variations in pose, facial expression, lighting conditions, etc.

matching methods use pre-stored face templates to judge if an image is a face; appearance-based methods learn face models from a set of representative training face images to perform detection. In general, appearance-based methods had been showing superior performance to the others, thanks to the rapid growing computation power and data storage.

In the following, we give a brief survey on the latest development in face detection techniques since the publication of Yang et al. (2002). As the title of this book suggests, the focus will be on boosting-based face detection schemes, which have evolved as the de-facto standard of face detection in real-world applications. Compared with other existing machine learning algorithms such as decision trees, the support vector machines, the neural networks, etc., the boosting algorithm is very simple yet extremely effective (Caruana and Niculescu-Mizil, 2006). Combined with efficient features and the cascade structure (Viola and Jones, 2001), boosting-based face detector can perform the detection task at very high accuracy and in real-time, making it very attractive for real-world applications. In addition, a boosting-based face detector can be further enhanced with post filters that employs other more sophisticated features and potentially more accurate learning methods, and the merit of the boosting-based detector still shines since it is capable to eliminate most of the negative examples very efficiently.

1.2   THE VIOLA-JONES FACE DETECTOR

If one were asked to name a single face detection algorithm that has the most impact in the 2000's, it will most likely be the seminal work by Viola and Jones (2001). The Viola-Jones face detector contains three main ideas that make it possible to build a successful face detector that can run *in real time*: the integral image, classifier learning with AdaBoost, and the attentional cascade structure.

1.2.1   THE INTEGRAL IMAGE

Integral image, also known as a summed area table, is an algorithm for quickly and efficiently computing the sum of values in a rectangle subset of a grid. It was first introduced to the computer graphics field by Crow (1984) for use in mipmaps. Viola and Jones applied the integral image for rapid computation of Haar-like features, as detailed below.

The integral image is constructed as follows:

$$ii(x, y) = \sum_{x' \leq x, y' \leq y} i(x', y'), \tag{1.1}$$

where $ii(x, y)$ is the integral image at pixel location (x, y) and $i(x', y')$ is the original image. The integral image can be constructed in one pass over the original image using the following recurrent formulas:

$$s(x, y) = s(x, y - 1) + i(x, y); \tag{1.2}$$
$$ii(x, y) = ii(x - 1, y) + s(x, y), \tag{1.3}$$

where $s(x, y)$ is the accumulated pixel values of row x, $s(x, -1) = 0$, and $ii(-1, y) = 0$.

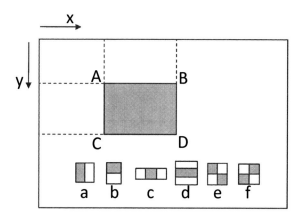

Figure 1.2: Illustration of the integral image and Haar-like rectangle features (a-f).

Using the integral image to compute the sum of any rectangular area is extremely efficient, as shown in Fig. 1.2. The sum of pixels in rectangle region $ABCD$ can be calculated as:

$$\sum_{(x,y)\in ABCD} i(x,y) = ii(D) + ii(A) - ii(B) - ii(C), \tag{1.4}$$

which only requires four array references.

The integral image can be used to compute simple Haar-like rectangular features, as shown in Fig. 1.2 (a-f). The features are defined as the (weighted) intensity difference between two to four rectangles. For instance, in feature (a), the feature value is the difference in average pixel value in the gray and white rectangles. Since the rectangles share corners, the computation of two rectangle features (a and b) requires six array references, the three rectangle features (c and d) requires eight array references, and the four rectangle features (e and f) requires nine array references.

1.2.2 ADABOOST LEARNING

Boosting is a method of finding a highly accurate classifier by combining many "weak" classifiers, each with moderate accuracy. An intuitive explanation of the boosting algorithm is shown in Fig. 1.3. The training examples are given their own weights, initialized as being all equal. The best weak classifier is then chosen. For those examples that are misclassified, their weights will be increased, and for those that are correctly classified, their weights will be reduced. The next weak classifier will be chosen with emphasis on the examples that were misclassified. Therefore, even though each weak classifier has limited accuracy, the combined classifier will still have very high performance. For an introduction on boosting, we refer the readers to Meir and Rätsch (2003) and Friedman et al. (1998).

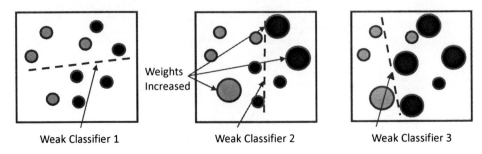

Weak Classifier 1   Weak Classifier 2   Weak Classifier 3

Figure 1.3: An intuitive explanation of boosting (Bishop and Viola, 2003). Red and blue circles represent examples in two categories. The size of each circle represents its weight in boosting. The final classifier is the linear combination of a series of weak classifiers.

The AdaBoost (Adaptive Boosting) algorithm is generally considered as the first step towards more practical boosting algorithms (Freund and Schapire, 1994, 1997). In this section, following Schapire and Singer (1999) and Friedman et al. (1998), we present a generalized version of AdaBoost algorithm, usually referred as *RealBoost*. It has been advocated in various works (Li et al., 2002; Bishop and Viola, 2003; Wu et al., 2004a; Mita et al., 2005) that RealBoost yields better performance than the original AdaBoost algorithm.

Consider a set of training examples as $S = \{(x_i, z_i), i = 1, \cdots, N\}$, where x_i belongs to a domain or instance space $\mathcal{X}$, and z_i belongs to a finite label space $\mathcal{Z}$. In binary classification problems, $\mathcal{Z} = \{0, 1\}$, where $z_i = 1$ for positive examples and $z_i = 0$ for negative examples. AdaBoost produces an additive model:

$$F^T(x) = \sum_{t=1}^{T} f_t(x), \tag{1.5}$$

to predict the label of an input example x, where $f_t(x)$ is a base function to be explained later, $F^T(x)$ is a real valued function in the form $F^T : \mathcal{X} \to \mathbb{R}$. The predicted label:

$$\hat{z}_i = \frac{1}{2} \left[\text{sign}(F^T(x_i)) + 1 \right], \tag{1.6}$$

where $\text{sign}(\cdot)$ is the sign function. From the statistical view of boosting (Friedman et al., 1998), AdaBoost algorithm fits an additive logistic regression model by using adaptive Newton updates for minimizing the expected exponential criterion:

$$L^T = \sum_{i=1}^{N} [I(z_i = 1) \exp\{-F^T(x_i)\} + I(z_i = 0) \exp\{F^T(x_i)\}], \tag{1.7}$$

where $I(\cdot)$ is the indicator function, such that $I(\text{true}) = 1$ and $I(\text{false}) = 0$.

The AdaBoost learning algorithm can be considered as to find the best additive base function $f_{t+1}(x)$ once $F^t(x)$ is given. For this purpose, we assume the base function pool $\{f(x)\}$ is in the form of confidence rated decision stumps. That is, a certain form of real feature value $h(x)$ is first extracted from $x, h : \mathcal{X} \to \mathbb{R}$. For instance, in the Viola-Jones face detector, $h(x)$ is the Haar-like features computed with integral image, as was shown in Fig. 1.2 (a-f). A decision threshold H divide the output of $h(x)$ into two subregions, u_1 and $u_2, u_1 \cup u_2 = \mathbb{R}$. The base function $f(x)$ is thus:

$$f(x) = \begin{cases} c_1, & \text{if } h(x) \in u_1; \\ c_2, & \text{if } h(x) \in u_2. \end{cases} \tag{1.8}$$

which is often referred as the stump classifier. Here c_1 and c_2 are called confidences. It is understandable that such stump classifiers usually have poor performance, thus the name "weak classifiers". It has been shown that as long as the weak classifier's classification accuracy is better than 50%, or random guess, boosting will improve the assembled classifier's accuracy (Freund and Schapire, 1994).

The optimal values of the confidence values can be derived as follows. For $j = 1, 2$, let

$$W_{+j} = \sum_i I(f(x_i) \in u_j) I(z_i = 1) \exp\{-F^t(x_i)\}$$

$$W_{-j} = \sum_i I(f(x_i) \in u_j) I(z_i = 0) \exp\{F^t(x_i)\}. \tag{1.9}$$

The target criterion can thus be written as:

$$\begin{aligned} L^{t+1} &= \sum_{i=1}^N [I(z_i = 1) \exp\{-F^t(x_i)\} \exp\{-f(x_i)\} + \\ & \quad I(z_i = 0) \exp\{F^T(x_i)\} \exp\{f(x_i)\}, \\ &= \sum_{j=1}^2 \left[W_{+j} e^{-c_j} + W_{-j} e^{c_j} \right]. \end{aligned} \tag{1.10}$$

Using standard calculus, we see L^{t+1} is minimized when

$$c_j = \frac{1}{2} \ln \left(\frac{W_{+j}}{W_{-j}} \right). \tag{1.11}$$

Plugging it into (1.10), we have:

$$L^{t+1} = 2 \sum_{j=1}^2 \sqrt{W_{+j} W_{-j}}. \tag{1.12}$$

Eq. (1.12) is referred as the Z score in (Schapire and Singer, 1999). In practice, at iteration $t + 1$, for every Haar-like feature $h(x)$, we find the optimal threshold H and confidence score c_1 and c_2 in order to minimize the Z score L^{t+1}. A simple pseudo code of the AdaBoost algorithm is shown in Fig. 1.4.

Input

- Training examples $S = \{(x_i, z_i), i = 1, \cdots, N\}$.
- T is the total number of weak classifiers to be trained.

Initialize

- Initialize example score $F^0(x_i) = \frac{1}{2} \ln \left(\frac{N_+}{N_-} \right)$, where N_+ and N_- are the number of positive and negative examples in the training data set.

Adaboost Learning
For $t = 1, \cdots, T$:

1. For each Haar-like feature $h(x)$ in the pool, find the optimal threshold H and confidence score c_1 and c_2 to minimize the Z score L^t (1.12).
2. Select the best feature with the minimum L^t as $f_t(x)$ in (1.5).
3. Update $F^t(x_i) = F^{t-1}(x_i) + f_t(x_i), i = 1, \cdots, N$,
4. Update $W_{+j}, W_{-j}, j = 1, 2$.

Output Final classifier $F^T(x)$.

Figure 1.4: Adaboost learning pseudo code.

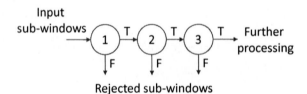

Figure 1.5: The attentional cascade. Each node is a boosted classifier.

1.2.3   THE ATTENTIONAL CASCADE STRUCTURE

Attentional cascade is a critical component in the Viola-Jones detector. The key insight is that smaller, and thus more efficient, boosted classifiers can be built which reject most of the negative sub-windows while keeping almost all the positive examples. Consequently, majority of the sub-windows will be rejected in early stages of the detector, making the detection process extremely efficient.

The overall process of classifying a sub-window thus forms a degenerate decision tree, which was called a "cascade" in (Viola and Jones, 2001). As shown in Fig. 1.5, the input sub-windows pass a series of nodes during detection. Each node is a boosted classifier, which will make a binary decision on whether the window will be kept for the next round or rejected immediately. The number of weak classifiers in the nodes usually increases as the number of nodes a sub-window passes. For instance, in (Viola and Jones, 2001), the first five nodes contain 1, 10, 25, 25, 50 weak classifiers, respectively. This is intuitive, since each node is trying to reject a certain amount of negative windows while keeping all the positive examples, and the task becomes harder at late stages. Having fewer weak classifiers at early stages also improves the speed of the detector.

The cascade structure also has an impact on the training process. Face detection is a rare event detection task. Consequently, there are usually billions of negative examples needed in order to train a high performance face detector. To handle the huge amount of negative training examples, Viola and Jones used a bootstrap process. That is, at each node, a threshold was manually chosen, and the partial classifier was used to scan the negative example set to find more unrejected negative examples for the training of the next node. Furthermore, each node is trained independently as if the previous nodes does not exist. One argument behind such a process is to force the addition of some nonlinearity in the training process, which could improve the overall performance. However, recent works showed that it is actually beneficial not to completely separate the training process of different nodes, as will be discussed in Section 1.3.2.

In Viola and Jones (2001), the attentional cascade is constructed manually. That is, the number of weak classifiers and the decision threshold for early rejection at each node are both specified manually. This is a non-trivial task. If the decision thresholds were set too aggressively, the final detector will be very fast, but the overall detection rate (percentage of true faces detected) may be hurt. On the other hand, if the decision thresholds were set very conservatively, most sub-windows will need to pass through many nodes, making the detector very slow. Combined with the limited computational resources available in early 2000's, it is no wonder that training a good face detector can take months of fine-tuning.

1.3 RECENT ADVANCES IN FACE DETECTION

Thanks to the rapid expansion in storage and computation resources, appearance-based methods have dominated the recent advances in face detection. The general practice is to collect a large set of face and non-face examples and adopt certain machine learning algorithms to learn a face model to perform classification. There are two key issues in this process: what features to extract and which learning algorithm to apply. In the following, we briefly review the recent advances in face detection based on these two key aspects.

1.3.1 FEATURE EXTRACTION

The Haar-like rectangular features as in Fig. 1.2 (a-f) are very efficient to compute due to the integral image technique and provide good performance for building frontal face detectors. However, these features have their limitations when handling non-frontal faces, faces under strong lighting variations, etc. In recent works, researchers have proposed numerous ways to improve the feature extraction process. We give a summary of these various algorithms in Table 1.1. More detailed explanation of the different methods are given below.

Table 1.1: Features for face/object detection.

Feature Type	Representative Works
Haar-like features and its variations	Haar-like features (Viola and Jones, 2001)
	Rotated Haar-like features (Lienhart and Maydt, 2002)
	Rectangular features with structure (Li et al., 2002; Jones and Viola, 2003)
	Haar-like features on motion filtered image (Jones et al., 2003)
Pixel-based features	Pixel pairs (Baluja et al., 2004)
	Control point set (Abramson and Steux, 2005)
Binarized features	Modified census transform (Fröba and Ernst, 2004)
	LBP features (Jin et al., 2004; Zhang et al., 2007b)
	Locally assembled binary feature (Yan et al., 2008)
Generic linear features	Anisotropic Gaussian filters (Meynet et al., 2007)
	LNMF (Chen et al., 2001)
	Generic linear features with KL boosting (Liu and Shum, 2003)
	RNDA (Wang and Ji, 2005)
Statistics-based features	Edge orientation histograms (Levi and Weiss, 2004; Dalal and Triggs, 2005) etc.
	Spectral histogram (Waring and Liu, 2005)
	Spatial histogram (LBP-based) (Zhang et al., 2006)
	HoG and LBP (Wang et al., 2009a)
	Region covariance (Tuzel et al., 2006)
Composite features	Joint Haar-like features (Mita et al., 2008)
	Sparse feature set (Huang et al., 2006)
Shape features	Boundary/contour fragments (Opelt et al., 2006; Shotton et al., 2005)
	Edgelet (Wu and Nevatia, 2005)
	Shapelet (Sabzmeydani and Mori, 2007)

A number of researchers proposed to extend the Haar-like features with more variations in the ways rectangle features are combined. For instance, as shown in Fig. 1.6, Lienhart and Maydt (2002) generalized the feature set of Viola and Jones (2001) by introducing 45 degree rotated rectangular features (a-d), and center-surround features (e-f). In order to compute the 45 degree rotated

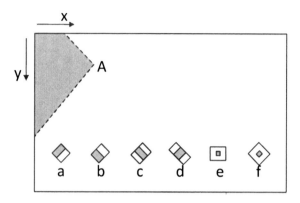

Figure 1.6: The rotated integral image/summed area table.

rectangular features, a new rotated summed area table was introduced as:

$$rii(x, y) = \sum_{x' \leq x, |y-y'| \leq x-x'} i(x', y').$$

(1.13)

As seen in Fig. 1.6, $rii(A)$ is essentially the sum of pixel intensities in the shaded area. The rotated summed area table can be calculated with two passes over all pixels.

In (Li et al., 2002; Li and Zhang, 2004), three types of features were defined in the detection sub-window, as shown in Fig. 1.7 (a). The rectangles are of flexible sizes $x \times y$, and they are at certain distances of (dx, dy) apart. The authors argued that these features can be non-symmetrical to cater to non-symmetrical characteristics of non-frontal faces. Jones and Viola (2003) also proposed a similar feature called diagonal filters, as shown in Fig. 1.7 (b). These diagonal filters can be computed with 16 array references to the integral image.

Jones et al. (2003) further extended the Haar-like feature set to work on motion filtered images for video-based pedestrian detection. Let the previous and current video frames be i_{t-1} and i_t. Five motion filters are defined as:

$$
\begin{aligned}
\Delta &= |i_t - i_{t-1}| \\
U &= |i_t - i_{t-1} \uparrow| \\
L &= |i_t - i_{t-1} \leftarrow| \\
R &= |i_t - i_{t-1} \rightarrow| \\
D &= |i_t - i_{t-1} \downarrow|
\end{aligned}
$$

where $\{\uparrow, \leftarrow, \rightarrow, \downarrow\}$ are image shift operators. $i_t \uparrow$ is i_t shifted up by one pixel. In addition to the regular rectangular features (Fig. 1.2) on these additional motion filtered images, Jones et al. added single box rectangular sum features and new features across two images. For instance:

$$f_i = r_i(\Delta) - r_i(S),$$

(1.14)

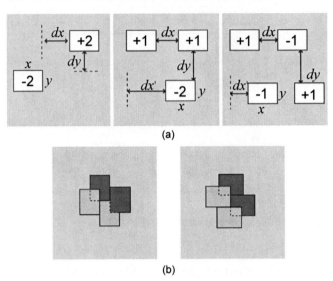

(a)

(b)

Figure 1.7: (a) Rectangular features with flexible sizes and distances introduced in (Li et al., 2002; Li and Zhang, 2004). (b) Diagonal filters in Jones and Viola (2003).

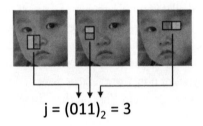

$$j = (011)_2 = 3$$

Figure 1.8: The joint Haar-like feature introduced in (Mita et al., 2005, 2008).

where $S \in \{U, L, R, D\}$ and $r_i(\cdot)$ is a single box rectangular sum within the detection window.

One must be careful that the construction of the motion filtered images $\{U, L, R, D\}$ is not scale invariant. That is, when detecting pedestrians at different scales, these filtered images need to be recomputed. This can be done by first constructing a pyramid of images for i_t at different scales and computing the filtered images at each level of the pyramid, as was done in Jones et al. (2003).

Mita et al. (2005, 2008) proposed joint Haar-like features, which is based on co-occurrence of multiple Haar-like features. The authors claimed that feature co-occurrence can better capture the characteristics of human faces, making it possible to construct a more powerful classifier. As shown in Fig. 1.8, the joint Haar-like feature uses a similar feature computation and thresholding scheme; however, only the binary outputs of the Haar-like features are concatenated into an index for 2^F possible combinations, where F is the number of combined features. To find distinctive feature co-occurrences with limited computational complexity, the suboptimal sequential forward selection scheme was used in (Mita et al., 2005). The number F was also heuristically limited to avoid statistical unreliability.

To some degree, the above joint Haar-like features resemble a CART tree, which was explored in Brubaker et al. (2005). It was shown that CART-tree-based weak classifiers improved results across various boosting algorithms with a small loss in speed. In another variation for improving the weak classifier, Wu et al. (2004a) proposed to use a single Haar-like feature, and equally bin the feature values into a histogram to be used in a RealBoost learning algorithm. Similar to the number F in the joint Haar-like features, the number of bins for the histogram is vital to the performance of the final detector. Wu et al. (2004a) proposed to use 64 bins. And in their later work (Huang et al., 2005), they specifically pointed out that too fine granularity of the histogram may cause overfitting, and they suggested to use fine granularity in the first few layers of the cascade, and coarse granularity in latter layers. Another interesting recent work is Xiao et al. (2007), where the authors proposed a new weak classifier called Bayesian stump. Bayesian stump is also a histogram-based weak classifier; however, the split thresholds of the Bayesian stump are derived from iterative split and merge operations instead of being at equal distances and fixed. Experimental results showed that such a flexible multi-split thresholding scheme is effective in improving the detector's performance. In Chapter 2 of this book, we will describe a similar scheme called fat stumps, which is more coherent with the RealBoost framework introduced before.

Another limitation of the original Haar-like feature set is its lack of robustness in handling faces under extreme lighting conditions, despite that the Haar features are usually normalized by the test windows' intensity covariance (Viola and Jones, 2001). In (Fröba and Ernst, 2004) a modified census transform was adopted to generate illumination-insensitive features for face detection. On each pixel's 3×3 neighborhood, the authors applied a modified census transform that compares the neighborhood pixels with their intensity mean. The results are concatenated into an index number representing the pixel's local structure. During boosting, the weak classifiers are constructed by examining the distributions of the index numbers for the pixels. Another well-known feature set robust to illumination variations is the local binary patterns (LBP) (Ojala et al., 2002), which have been very effective for face recognition tasks (Ahonen et al., 2004; Zhang et al., 2004). In (Jin et al., 2004; Zhang et al., 2007b), LBP was applied for face detection tasks under a Bayesian and a boosting framework, respectively. More recently, inspired by LBP, Yan et al. (2008) proposed locally assembled binary feature, which showed great performance on standard face detection data sets.

To explore possibilities to further improve performance, more and more complex features were proposed in the literature. For instance, Liu and Shum (2003) studied generic linear features, which is defined by a mapping function $\phi() : \mathbb{R}^d \rightarrow \mathbb{R}^1$, where d is the size of the test patch. For linear features, $\phi(x) = \phi^T x, \phi \in \mathbb{R}^d$. The classification function is in the following form:

$$F^T(x) = \text{sign} \left[\sum_t^T \lambda_t(\phi_t^T x) \right], \tag{1.15}$$

where $\lambda_t()$ are $\mathbb{R} \to \mathbb{R}$ discriminating functions, such as the conventional stump classifiers in Ad-aBoost. $F^T(x)$ shall be 1 for positive examples and -1 for negative examples. Note the Haar-like feature set is a subset of linear features. Another example is the anisotropic Gaussian filters in (Meynet et al., 2007). In (Chen et al., 2001), the linear features were constructed by pre-learning them using local non-negative matrix factorization (LNMF), which is still sub-optimal. Instead, Liu and Shum (2003) proposed to search for the linear features by examining the Kullback-Leibler (KL) divergence of the positive and negative histograms projected on the feature during boosting (hence the name Kullback-Leibler boosting). In Wang and Ji (2005), the authors proposed to apply Fisher discriminant analysis and more generally recursive nonparametric discriminant analysis (RNDA) to find the linear projections ϕ_t. Linear projection features are very powerful features. The selected features shown in (Liu and Shum, 2003) and (Wang and Ji, 2005) were like face templates. They may significantly improve the convergence speed of the boosting classifier at early stages. However, caution must be taken to avoid overfitting if these features are to be used at the later stages of learning. In addition, the computational load of linear features are generally much higher than the traditional Haar-like features. Oppositely, Baluja et al. (2004) proposed to use simple pixel pairs as features, and Abramson and Steux (2005) proposed to use the relative values of a set of control points as features. Such pixel-based feature can be computed even faster than the Haar-like features; however, their discrimination power is generally insufficient to build high performance detectors.

Another popular complex feature for face/object detection is based on regional statistics such as histograms. Levi and Weiss (2004) proposed local edge orientation histograms, which computes the histogram of edges orientations in subregions of the test windows. These features are then selected by an AdaBoost algorithm to build the detector. The orientation histogram is largely invariant to global illumination changes, and it is capable of capturing geometric properties of faces that are difficult to capture with linear edge filters such as Haar-like features. However, similar to motion filters, edge-based histogram features are not scale invariant, hence one must first scale the test images to form a pyramid to make the local edge orientation histograms features reliable. Later, Dalal and Triggs (2005) proposed a similar scheme called histogram of oriented gradients (HoG), which became a very popular feature for human/pedestrian detection (Zhu et al., 2006; Grabner and Bischof, 2006; Suard et al., 2006; Laptev, 2006; Enzweiler and Gavrila, 2009). In (Waring and Liu, 2005), the authors proposed spectral histogram features, which adopts a broader set of filters before collecting the histogram features, including gradient filters, Laplacian of Gaussian filters and Gabor filters. Compared with (Levi and Weiss, 2004), the histogram features in (Waring and Liu, 2005) were based on the whole testing window rather than local regions, and support vector machines (SVMs) were used for classification. Zhang et al. (2006) proposed another histogram-based feature called spatial histograms, which is based on local statistics of LBP. HoG and LBP were also combined in (Wang et al., 2009a), which achieved excellent performance on human detection with partial occlusion handling. Region covariance was another statistics-based feature, proposed by Tuzel et al. (2006) for generic object detection and texture classification tasks. Instead of using histograms, they compute the covariance matrices among the color channels and gradient images. Regional covariance features can also be efficiently computed using integral images.

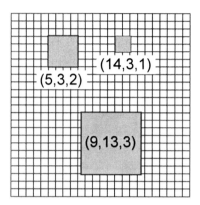

Figure 1.9: The sparse feature set in granular space introduced in Huang et al. (2006, 2007a).

Huang et al. (2006, 2007a) proposed a sparse feature set in order to strengthen the features' discrimination power without incurring too much additional computational cost. Each sparse feature can be represented as:

$$f(x) = \sum_i \alpha_i p_i(x; u, v, s), \alpha_i \in \{-1, +1\} \tag{1.16}$$

where x is an image patch, and p_i is a granule of the sparse feature. A granule is specified by 3 parameters: horizontal offset u, vertical offset v and scale s. For instance, as shown in Fig. 1.9, $p_i(x; 5, 3, 2)$ is a granule with top-left corner $(5,3)$, and scale $2^2 = 4$, and $p_i(x; 9, 13, 3)$ is a granule with top-left corner $(9,13)$, and scale $2^3 = 8$. Granules can be computed efficiently using pre-constructed image pyramids, or through the integral image. In (Huang et al., 2006), the maximum number of granules in a single sparse feature is 8. Since the total number of granules is large, the search space is very large and exhaustive search is infeasible. The authors proposed a heuristic search scheme, where granules are added to a sparse feature one-by-one, with an expansion operator that removes, refines and adds granules to a partially selected sparse feature. To reduce the computation, the authors further conducted multi-scaled search, which uses a small set of training examples to evaluate all features first and rejects those that are unlikely to be good. The performance of the multi-view face detector trained in (Huang et al., 2006) using sparse features was very good.

As new features are composed in seeking the best discrimination power, the feature pool becomes larger and larger, which creates new challenges in the feature selection process. A number of recent works have attempted to address this issue. For instance, Yuan et al. (2008) proposed to discover compositional features using the classic frequent item-set mining scheme in data mining. Instead of using the raw feature values, they assume a collection of induced binary features (e.g., decision stumps with known thresholds) are already available. By partitioning the feature space into sub-regions through these binary features, the training examples can be indexed by the sub-regions they are located. The algorithm then searches for a small subset of compositional features that are both frequent to have statistical significance and accurate to be useful for label prediction. The final classifier is then learned based on the selected subset of compositional features through AdaBoost. In Han et al. (2008), the authors first established an analogue between compositional feature selection and generative image segmentation, and they applied the Swendsen-Wang Cut algorithm to generate n-partitions for the individual feature set, where each subset of the partition corresponds to a compositional feature. This algorithm re-runs for every weak classifier selected by the AdaBoost learning framework. On a person detection task tested, the composite features showed significant improvement, especially when the individual features were very weak (e.g., Haar-like features).

In some applications such as object tracking, even if the number of possible features is not extensive, an exhaustive feature selection is still impractical due to computational constraints. In (Liu and Yu, 2007), the authors proposed a gradient-based feature selection scheme for online boosting with primary applications in person detection and tracking. Their work iteratively updates each feature using a gradient descent algorithm, by minimizing the weighted least square error between the estimated feature response and the true label. This is particularly attractive for tracking and updating schemes such as (Grabner and Bischof, 2006), where at any time instance, the object's appearance is already represented by a boosted classifier learned from previous frames. Assuming there is no dramatic change in the appearance, the gradient-descent-based algorithm can refine the features in a very efficient manner.

There have also been many features that were proposed to model the shape of the objects. For instance, Opelt et al. (2006) composed multiple boundary fragments to weak classifiers and formed a strong "boundary-fragment-model" detector using boosting. They ensure the feasibility of the feature selection process by limiting the number of boundary fragments to 2-3 for each weak classifier. Shotton et al. (2005) learned their object detectors with a boosting algorithm and their feature set consisted of a randomly chosen dictionary of contour fragments. A very similar edgelet feature was proposed in (Wu and Nevatia, 2005), and it was used to learn human body part detectors in order to handle multiple, partially occluded humans. In (Sabzmeydani and Mori, 2007), shapelet features focusing on local regions of the image were built from low-level gradient information using AdaBoost for pedestrian detection. An interesting side benefit of having contour/edgelet features is that object detection and object segmentation can be performed jointly, such as the work in (Wu and Nevatia, 2007b) and (Gao et al., 2009).

1.3.2 VARIATIONS OF THE BOOSTING LEARNING ALGORITHM

In addition to exploring better features, another venue to improve a detector's performance is through improving the boosting learning algorithm, particularly under the cascade decision structure. It turns out that there are many directions to be explored, thanks to the cascade structure that is required to run the detector efficiently. We first summarize various challenges and the proposed solutions in Table 1.2 and present them in more detail as below.

In the original face detection paper by Viola and Jones (2001), the standard AdaBoost algorithm (Freund and Schapire, 1994) was adopted. In a number of follow-up works (Li et al., 2002; Bishop and Viola, 2003; Wu et al., 2004a; Mita et al., 2005), researchers advocated the use of Real-Boost, which was explained in detail in Section 1.2.2. Both Lienhart et al. (2002) and Brubaker et al. (2005) compared three boosting algorithms: AdaBoost, RealBoost and GentleBoost, though they reach different conclusions as the former recommended GentleBoost while the latter showed Real-Boost works slightly better when combined with CART-based weak classifiers. In this book, we will adopt RealBoost with logistic weight (Collins et al., 2002), which will be detailed in Chapter 2. In the following, we describe a number of recent works on boosting learning for face/object detection, with emphasis on adapting to the cascade structure, the training speed, multi-view face detection, etc.

In (Li et al., 2002; Li and Zhang, 2004), the authors proposed FloatBoost, which attempted to overcome the monotonicity problem of the sequential AdaBoost Learning. Specifically, AdaBoost is a sequential forward search procedure using a greedy selection strategy, which may be suboptimal. FloatBoost incorporates the idea of floating search (Pudil et al., 1994) into AdaBoost, which not only add features during training, but also backtrack and examine the already selected features to remove those that are least significant. The authors claimed that FloatBoost usually needs fewer weak classifiers than AdaBoost to achieve a given objective. Jang and Kim (2008) proposed to used evolutionary algorithms to minimize the number of classifiers without degrading the detection accuracy. They showed that such an algorithm can reduce the total number of weak classifiers by over 40%. Note, in practice, only the first few nodes are critical to the detection speed since most testing windows are rejected by the first few weak classifiers in a cascade architecture.

As mentioned in 1.2.3, Viola and Jones (2001) trained each node independently. A number of follow-up works showed that there is indeed information in the results from the previous nodes, and it is best to reuse them instead of starting from scratch at each new node. For instance, in (Xiao et al., 2003), the authors proposed to use a "chain" structure to integrate historical knowledge into successive boosting learning. At each node, the existing partial classifier is used as a prefix classifier for further training. Boosting chain learning can thus be regarded as a variant of AdaBoost learning with similar generalization performance and error bound. In (Wu et al., 2004a), the authors proposed the so-called nesting-structured cascade. Instead of taking the existing partial classifier as a prefix, they took the confidence output of the partial classifier and used it as a feature to build the first weak classifier. Both paper demonstrated better detection performance than the original Viola-Jones face detector.

One critical challenge in training a cascade face detector is how to set the thresholds for the intermediate nodes. This issue has inspired a lot of works in the literature. First, Viola and Jones (2002) observed that the goal of the early stages of the cascade is mostly to retain a very high detection rate, while accepting modest false positive rates if necessary. They proposed a new scheme called asymmet-

Table 1.2: Face/object detection schemes to address challenges in boosting learning.

Challenges	Representative Works
General boosting schemes	AdaBoost (Viola and Jones, 2001)
	RealBoost (Li et al., 2002; Bishop and Viola, 2003; Wu et al., 2004a; Mita et al., 2005)
	GentleBoost (Lienhart et al., 2002; Brubaker et al., 2005)
	FloatBoost (Li et al., 2002)
Reuse previous nodes' results	Boosting chain (Xiao et al., 2003)
	Nested cascade (Wu et al., 2004a)
Introduce asymmetry	Asymmetric Boosting (Viola and Jones, 2002; Pham and Cham, 2007a; Masnadi-Shirazi and Vasconcelos, 2007a)
	Linear asymmetric classifier (Wu et al., 2005)
Set intermediate thresholds during training	Fixed node performance (Lienhart et al., 2002)
	WaldBoost (Sochman and Matas, 2005)
	Based on validation data (Brubaker et al., 2005)
	Exponential curve (Xiao et al., 2007)
Set intermediate thresholds after training	Greedy search (Luo, 2005)
	Soft cascade (Bourdev and Brandt, 2005)
	Multiple instance pruning (Zhang and Viola, 2007) (Chapter 2)
Speed up training	Greedy search in feature space (McCane and Novins, 2003)
	Random feature subset (Brubaker et al., 2005)
	Forward feature selection (Wu et al., 2004b)
	Use feature statistics (Pham and Cham, 2007b)
Speed up testing	Reduce number of weak classifiers (Li et al., 2002; Jang and Kim, 2008)
	Feature centric evaluation (Schneiderman, 2004b; Yan et al., 2008)
	Caching/selective attention (Pham and Cham, 2005) etc.
Multiview face detection	Parallel cascade (Wu et al., 2004a)
	Pyramid structure (Li et al., 2002)
	Decision tree (Jones and Viola, 2003; Fröba and Ernst, 2003)
	Vector valued boosting (Huang et al., 2005; Lin and Liu, 2005)
Learn without subcategory labels	Cluster and then train (Seemann et al., 2006)
	Exemplar-based learning (Shan et al., 2006)
	Probabilistic boosting tree (Tu, 2005)
	Cluster with selected features (Wu and Nevatia, 2007a)
	Multiple classifier/category boosting (Kim and Cipolla, 2008; Babenko et al., 2008) and Chapter 3

ric AdaBoost, which artificially increase the weights on positive examples in each round of AdaBoost such that the error criterion biases towards having low false negative rates. In (Pham and Cham, 2007a), the authors extended the above work and sought to balance the skewness of labels presented to each weak classifiers, so that they are trained more equally. Masnadi-Shirazi and Vasconcelos (2007a) further proposed a more rigorous form of asymmetric boosting based on the statistical interpretation of boosting (Friedman et al., 1998) with an extension of the boosting loss. Namely, the exponential cost criterion in Eq. (1.7) is rewritten as:

$$L^T = \sum_{i=1}^{N} [I(z_i = 1) \exp\{-C_1 F^T(x_i)\} + $$
$$I(z_i = 0) \exp\{C_0 F^T(x_i)\}], \qquad (1.17)$$

where C_1 and C_0 are the weights associated with positive and negative examples, respectively. Masnadi-Shirazi and Vasconcelos (2007a) minimized the above criterion following the AnyBoost framework in (Mason et al., 2000). They were able to build a detector with very high detection rate (Masnadi-Shirazi and Vasconcelos, 2007b), though the performance of the detector deteriorates very quickly when the required false positive rate is low.

Wu et al. (2005, 2008) proposed to decouple the problems of feature selection and ensemble classifier design in order to introduce asymmetry. They first applied the forward feature selection algorithm to select a set of features, and then formed the ensemble classifier by voting among the selected features through a linear asymmetric classifier (LAC). The LAC is the optimal linear classifier under the assumption that the linear projection of the features for positive examples follows a Gaussian distribution, and that for negative examples is symmetric. Mathematically, LAC has a similar form as the well-known Fisher discriminant analysis (FDA) (Duda et al., 2001), except that only the covariance matrix of the positive feature projections are considered in LAC. In practice, their performance are also similar. Applying LAC or FDA on a set of features pre-selected by AdaBoost is equivalent to readjust the confidence values of the AdaBoost learning (Eq. (1.11)). Since at each node of the cascade, the AdaBoost learning usually has not converged before moving to the next node, readjusting these confidence values could provide better performance for that node. However, when the full cascade classifier is considered, the performance improvement over AdaBoost diminished. Wu et al. attributed the phenomenon to the booststrapping step and the post processing step, which also have significant effects on the cascade's performance.

With or without asymmetric boosting/learning, at the end of each cascade node, a threshold still has to be set in order to allow the early rejection of negative examples. These node thresholds reflect a tradeoff between detection quality and speed. If they are set too aggressively, the final detector will be fast, but the detection rate may drop. On the other hand, if the thresholds are set conservatively, many negative examples will pass the early nodes, making the detector slow. In early works, the rejection thresholds were often set in very ad hoc manners. For instance, Viola and Jones (2001) attempted to reject zero positive examples until this become impossible and then reluctantly gave up on one positive example at a time. Huge amount of manual tuning is thus required to find a classifier with good balance between quality and speed, which is very inefficient. Lienhart et al. (2002) instead built the cascade targeting each node to have 0.1% false negative rate and 50% rejection rate for the negative examples. Such a scheme is simple to implement, though no speed guarantee can be made about the final detector.

In (Sochman and Matas, 2005), the authors proposed to use a ratio test to determine the rejection thresholds. Specifically, the authors viewed the cascade detector as a sequential decision-making problem. A sequential decision-making theory had been developed by Wald (Wald, 1947), which proved that the solution to minimizing the expected evaluation time for a sequential decision-making problem is the sequential probability ratio test. Sochman and Matas abandoned the notion of nodes, and set rejection threshold after each weak classifier. They then approximated the joint likelihood ratio of all the weak classifiers between negative and positive examples with the likelihood ratio of the partial scores, in which case the algorithm simplified to be rejecting a test example if the likelihood ratio at its partial score value is greater than $\frac{1}{\alpha}$, where α is the false negative rate of the entire cascade. Brubaker et al. (2005) proposed another fully automatic algorithm for setting the intermediate thresholds during training. Given the target detection and false positive rates, their algorithm used the empirical results on validation data to estimate the probability that the cascade will meet the goal criteria. Since a reasonable goal is not known a priori, the algorithm adjusts its cost function depending on the attainability of the goal based on cost prediction. In (Xiao et al., 2007), a dynamic cascade was proposed, which assumes that the false negative rate of the nodes changes exponentially in each stage, following an idea in (Bourdev and Brandt, 2005). The approach is simple and ad hoc, though it appears to work reasonably well.

Setting intermediate thresholds during training is a specific scheme to handle a huge amount of negative examples during boosting training. Such a step is unnecessary in AdaBoost, at least according to its theoretical derivation. Recent development of boosting-based face detector training have shifted toward approaches where these intermediate thresholds are not set during training, but rather done until the whole classifier has been learnt. For instance, Luo (2005) assumed that a cascade of classifiers is already designed, and proposed an optimization algorithm to adjust the intermediate thresholds. It represents each individual node with a uniform abstraction model with parameters (e.g., the rejection threshold) controlling the tradeoff between the detection rate and the false alarm rate (percentage of detected faces that are non-face). It then uses a greedy search strategy to adjust the parameters such that the slope of the logarithm scale ROC curves of all the nodes are equal. One issue in such a scheme is that the ROC curves of the nodes are dependent on changes in thresholds of any earlier nodes; hence, the greedy search scheme can at best be an approximation. Bourdev and Brandt (2005) instead proposed a heuristic approach to use a parameterized exponential curve to set the intermediate nodes' detection targets, called a "rejection distribution vector". By adjusting the parameters of the exponential curve, different tradeoffs can be made between speed and quality. Perhaps a particular family of curves is more palatable, but it is still arbitrary and non-optimal. In Chapter 2, we will present a more principled data-driven scheme for setting intermediate thresholds, which is based on the idea of multiple instance learning (Nowlan and Platt, 1995).

The remaining issue is how to train a cascade detector with billions of examples without explicitly setting the intermediate thresholds. In (Bourdev and Brandt, 2005), the authors proposed a scheme that starts with a small set of training examples, and adds to it new samples at each stage that the current classifier misclassifies. The number of new non-faces to be added at each training cycle affects the focus of AdaBoost during training. If the number is too large, AdaBoost may not be able to catch up and the false positive rate will be high. If the number is too small, the cascade may contain too many weak classifiers in order to reach a reasonable false positive rate. In addition, later stages of the training will be slow due to the increasing number of negative examples, since none of them will be removed during the process. Chapter 2 of this book will describe a sampling-based scheme that completely addresses the issue of large data set training for boosting approaches.

Training a face detector is a very time-consuming task. In early works, due to the limited computing resources, it could easily take months and lots of manual tuning to train a high quality face detector. The main bottleneck is at the feature selection stage, where hundreds of thousands of Haar features will need to be tested at each iteration. A number of papers has been published to speed up the feature selection process. For instance, McCane and Novins (2003) proposed a discrete downhill search scheme to limit the number of features compared during feature selection. Such a greedy search strategy offered a 300–400 fold speed up in training, though the false positive rate of the resultant detector increased by almost a factor of 2. Brubaker et al. (2005) studied various filter schemes to reduce the size of the feature pool, and they showed that randomly selecting a subset of features at each iteration for feature selection appears to work reasonably well. Wu et al. (2004b) proposed a cascade learning algorithm based on forward feature selection (Webb, 1999), which is two orders of magnitude faster than the traditional approaches. The idea is to first train a set of weak classifiers that satisfy the maximum false positive rate requirement of the entire detector. During feature selection, these weak classifiers are added one by one, each making the largest improvement to the ensemble performance. Weighting of the weak classifiers can be conducted after the feature selection step. Pham and Cham (2007b) presented another fast method to train and select Haar features. It treated the training examples as high dimensional random vectors and kept the first and second order statistics to build classifiers from features. The time complexity of the method is linear to the total number of examples and the total number of Haar features. Both (Wu et al., 2004b) and (Pham and Cham, 2007b) reported experimental results demonstrating better ROC curve performance than the traditional AdaBoost approach, though it appears unlikely that they can also outperform t

Various efforts have also been made to improve the detector's test speed. For instance, in the sparse feature set in (Huang et al., 2006), the authors limited the granules to be in square shape, which is very efficient to compute in both software and hardware through building pyramids for the test image. For HoG and similar gradient-histogram-based features, the integral histogram approach (Porikli, 2005) was often adopted for faster detection. Schneiderman (2004b) designed a feature-centric cascade to speed up the detection. The idea is to pre-compute a set of feature values over a regular grid in the image, so that all the test windows can use their corresponding feature values for the first stage of the detection cascade. Since many feature values are shared by multiple windows, significant gains in speed can be achieved. A similar approach was deployed in (Yan et al., 2008) to speed up their locally assembled binary-feature-based detector. In (Pham and Cham, 2005), the authors proposed a scheme to improve the detection speed on quasi-repetitive inputs, such as the video input during videoconferencing. The idea is to cache a set of image exemplars, each induces its own discriminant subspace. Given a new video frame, the algorithm quickly searches through the exemplar database indexed with an online version of tree-structured vector quantization, S-tree (Campos and Carpenter, 2001). If a similar exemplar is found, the face detector will be skipped and the previously detected object states will be reused. This results in about 5-fold improvement in detection speed. Similar amount of speed-up can also be achieve through selective attention, such as those based on motion, skin color, background modeling and subtraction, etc.

As shown in Fig. 1.1, in real-world images, faces have significant variations in orientation, pose, facial expression, lighting conditions, etc. A single cascade with Haar features has proven to work very well with frontal or near-frontal face detection tasks. However, extending the algorithm to multi-pose/multi-view face detection is not straightforward. If faces with all pose/orientation variations are trained in a single classifier, the results are usually sub-optimal. To this end, researchers have proposed numerous schemes to combat the issue, most of them following the "divide and conquer" strategy.

Fig. 1.10 showed a number of detector structures for multiview face detection. Among these structures, the most straightforward one is Fig. 1.10(a), the parallel cascade, by Wu et al. (2004a). An individual classifier is learned for each view. A test window is passed to all the classifiers. After a few nodes, one cascade with the highest score will finish the classification and make the decision. This simple structure could achieve rather good performance, though its running speed is generally slow, and the correlation between faces of different views could have been better exploited. Li et al. (2002) used a pyramid structure to handle the task, as shown in Fig. 1.10(b). The detector pyramid consists of 3 levels. The first level of the pyramid works on faces at all poses; the second level detects faces between $-90°$ and $-30°$ (left profile), between $-30°$ and $30°$ (frontal), and between $30°$ and $90°$ (right profile), respectively; the third level detects faces at 7 finer angles. Once a test window passes one level of the detector, it will be passed to all the children nodes for further decision. This design is more efficient than the parallel cascade structure, but it still has room to improve.

Fig. 1.10(c) and (d) showed two decision tree structures for multiview face detection. In (Jones and Viola, 2003), the authors proposed to first use a pose estimator to predict the face pose of a test window. Given the predicted pose, a cascade for that pose will be invoked to make the final decision. A decision tree was adopted for pose estimation, which resulted in the detector structure in Fig. 1.10(c). With this structure, a test window will only run through a single cascade once its pose has been estimated, thus the detector is very efficient. Fröba and Ernst (2003) had a similar tree structure (Fig. 1.10(d)) for frontal face detection at different orientations, except that their early nodes were able to perform rejection to further improve speed. However, pose/orientation estimation is a non-trivial task, and it can have many errors. If a profile face is misclassified as frontal, it may

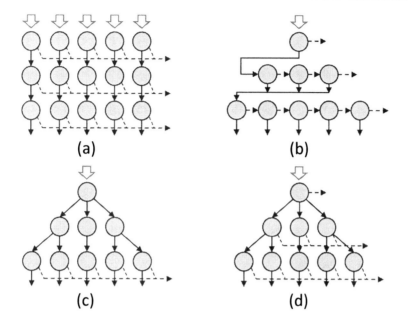

Figure 1.10: Various detector structures for multiview face detection. Each circle represents a strong classifier. The solid arrows are pass route, and the dashed arrows are reject route. (a) Parallel cascade (Wu et al., 2004a). (b) Detector-pyramid (Li et al., 2002). (c) Decision tree I (Jones and Viola, 2003). (d) Decision tree II (Fröba and Ernst, 2003; Huang et al., 2005; Lin and Liu, 2005). Note in (d) the early nodes are all able to perform rejection in order to speed up the detection. In addition, in (Huang et al., 2005; Lin and Liu, 2005) the selection of the pass route for a branching node is non-exclusive.

never be detected by the frontal face cascade. Huang et al. (2005, 2007a) and Lin and Liu (2005) independently proposed a very similar solution to this issue, which were named vector boosting and multiclass Bhattacharyya boost (MBHBoost), respectively. The idea is to have vector valued output for each weak classifier, which allows an example to be passed into multiple subcategory classifiers during testing (Fig. 1.10(d)), and the final results are fused from the vector output. Such a soft branching scheme can greatly reduce the risk of misclassification during testing. Another interesting idea in (Huang et al., 2005; Lin and Liu, 2005) was to have all the subcategory classifiers share the same features. Namely, at each iteration, only one feature is chosen to construct a weak classifier with vector output, effectively sharing the feature among all the subcategories. Sharing features among multiple classifiers had been shown as a successful idea to reduce the computational and sample complexity when multiple classifiers are trained jointly (Torralba et al., 2004).

Vector boosting and MBHBoost solved the issue of misclassification in pose estimation during testing. During training, they still used faces manually labeled with pose information to learn the multiview detector. However, for certain object classes such as pedestrians or cars, an agreeable manual pose labeling scheme is often unavailable. Seemann et al. (2006) extended the implicit shape model in (Leibe et al., 2005) to explicitly handle and estimate viewpoints and articulations of an object category. The training examples were first clustered, with each cluster representing one articulation and viewpoint. Separate models were then trained for each cluster for classification. Shan et al. (2006) proposed an exemplar-based categorization scheme for multiview object detection. At each round of boosting learning, the algorithm selects not only a feature to construct a weak classifier, but also a set of exemplars to guide the learning to focus on different views of the object. Tu (2005) proposed a probabilistic boosting tree, which embeds clustering in the learning phase. At each tree node, a strong AdaBoost-based classifier is built. The output of the AdaBoost classifier is used to compute the posterior probabilities of the examples, which are used to split the data into two clusters. In some sense, the traditional boosting cascade can be viewed as a special case of the boosting tree, where all the positive examples are pushed into one of the child node. The performance of boosting tree on multiview object detection is uncertain due to the limited experimental results provided in the paper. In (Wu and Nevatia, 2007a), a similar boosted tree algorithm was proposed. Instead of performing clustering before boosting learning or using posterior probabilities, they showed that by using the previously selected features for clustering, the learning algorithm converges faster and achieves better results.

Some recent works went one step further and did not maintain a fixed subcategory label for the training examples any more. For instance, Kim and Cipolla (2008) proposed an algorithm called multiple classifier boosting, which is a straightforward extension of the multiple instance boosting approach in Viola et al. (2005) (also see Chapter 3). In this approach, the training examples no longer have a fixed subcategory label. A set of likelihood values were maintained for each example, which describe the probability of it belonging to the subcategories during training. These likelihood values are combined to compute the probability of the example being a positive example. The learning algorithm then maximizes the overall probability of all examples in the training data set. Babenko et al. (2008) independently developed a very similar scheme they called multi-pose learning, and further combined it with multiple instance learning in a unified framework. One limitation of the above approaches is that the formulation requires a line search at each weak classifier to find the optimal weights, which makes it slow to train and hard to deploy feature sharing (Torralba et al., 2004). In Chapter 3, we will present an algorithm called winner-take-all multiple category boosting, which is more suitable for learning multiview detectors with huge amount of training data.

1.3.3   OTHER LEARNING SCHEMES

As reviewed in the previous section, the seminal work by Viola and Jones (2001) has inspired a lot of research applying the boosting cascade for face detection. Nevertheless, there were still a few papers approaching the problem in different ways, some providing very competitive performances. These works are summarized in Table 1.3. Again, we will only focus on works not covered in (Yang et al., 2002).

Table 1.3: Other schemes for face/object detection (since (Yang et al., 2002)).

General Approach	Representative Works
Template matching	Antiface (Keren et al., 2001)
Bayesian	Bayesian discriminating features (Liu and Shum, 2003)
SVM – speed up	Reduced set vectors and approximation (Romdhani et al., 2001; Rätsch et al., 2004)
	Resolution-based SVM cascade (Heisele et al., 2003)
SVM – multiview face detection	SVR-based pose estimator (Li et al., 2000)
	SVR fusion of multiple SVMs (Yan et al., 2001)
	Cascade and bagging (Wang and Ji, 2004)
	Local and global kernels (Hotta, 2007)
Neural networks	Constrained generative model (Féraud et al., 2001)
	Convolutional neural network (Garcia and Delakis, 2004; Osadchy et al., 2004)
Part-based approaches	Wavelet localized parts (Schneiderman and Kanade, 2004; Schneiderman, 2004a)
	SVM component detectors adaptively trained (Heisele et al., 2007)
	Overlapping part detectors (Mikolajczyk et al., 2004)

Keren et al. (2001) proposed Antifaces, a multi-template scheme for detecting arbitrary objects including faces in images. The core idea is very similar to the cascade structure in (Viola and Jones, 2001), which uses a set of sequential classifiers to detect faces and rejects non-faces fast. Each classifier, referred as a "detector" in (Keren et al., 2001), is a template image obtained through constrained optimization, where the inner product of the template with the example images are minimized, and the later templates are independent to the previous ones. Interestingly, in this approach, negative images were modeled by a Boltzmann distribution and assumed to be smooth, thus none is needed during template construction.

Liu (2003) presented a Bayesian discriminating features method for frontal face detection. The face class was modeled as a multivariate normal distribution. A subset of the nonfaces that lie closest to the face class was then selected based on the face class model and also modeled with a multivariate normal distribution. The final face/nonface decision was made by a Bayesian classifier. Since only the nonfaces closest to the face class were modeled, the majority of the nonfaces were ignored during the classification. This was inspired by the concept of support vector machines (SVMs) (Cristianini and Shawe-Taylor, 2000), where only a subset of the training examples (the support vectors) were used to define the final classifier.

SVMs are known as maximum margin classifiers, as they simultaneously minimize the empirical classification error and maximize the geometric margin. Due to their superior performance in general machine learning problems, they have also become a very successful approach for face detection (Osuna et al., 1997; Heisele et al., 2000). However, the speed of SVM-based face detectors was generally slow. Various schemes have since been proposed to speed up the process. For instance, Romdhani et al. (2001) proposed to compute a set of reduced set vectors from the original support vectors. These reduced set vectors are then tested against the test example sequentially, making early rejections possible. Later, Rätsch et al. (2004) further improved the speed by approximating the reduced set vectors with rectangle groups, which gained another 6-fold speedup. Heisele et al. (2003) instead used a hierarchy of SVM classifiers with different resolutions in order to speed up the overall system. The early classifiers are at low resolution, say, 3×3 and 5×5 pixels, which can be computed very efficiently to prune negative examples.

Multiview face detection has also been explored with SVM-based classifiers. Li et al. (2000) proposed a multiview face detector similar to the approach in (Rowley et al., 1997; Jones and Viola, 2003). They first constructed a face pose estimator using support vector regression (SVR), then trained separate face detectors for each face pose. Yan et al. (2001) instead executed multiple SVMs first, and then applied an SVR to fuse the results and generate the face pose. This method is slower, but it has lower risk of assigning a face to the wrong pose SVM and causing misclassification. Wang and Ji (2004) remarked that in the real world the face poses may vary greatly and many SVMs are needed. They proposed an approach to combine cascade and bagging for multiview face detection. Namely, a cascade of SVMs were first trained through bootstrapping. The remaining positive and negative examples were then randomly partitioned to train a set of SVMs, whose outputs were then combined through majority voting. Hotta (2007) used a single SVM for multiview face detection, and relied on the combination of local and global kernels for better performance. No experimental results were given in (Wang and Ji, 2004; Hotta, 2007) to compare the proposed methods with existing schemes on standard data sets; hence, it is unclear whether these latest SVM-based face detectors can outperform those learned through boosting.

Neural networks were another popular approach to build a face detector. Early representative methods included the detectors by Rowley et al. (1996) and Roth et al. (2000). Féraud et al. (2001) proposed an approach based on a neural network model called the constrained generative model (CGM). CGM is an autoassociative, fully connected multilayer perceptron (MLP) with three large layers of weights, trained to perform nonlinear dimensionality reduction in order to build a generative model for faces. Multiview face detection was achieved by measuring the reconstruction errors of multiple CGMs, combined via a conditional mixture and an MLP gate network. In (Garcia and Delakis, 2004), the authors proposed a face detection scheme based on a convolutional neural architecture. Compared with traditional feature-based approaches, convolutional neural network derives problem-specific feature extractors from the training examples automatically, without making any assumptions about the features to extract or the areas of the face patterns to analyze. Osadchy et al. (2004) proposed another convolutional-network-based approach, which was able to perform multiview face detection and facial pose estimation simultaneously. The idea is to train a convolutional neural network to map face images to points on a low dimensional face manifold parameterized by facial pose, and non-face images to points far away from the manifold. The detector was fast and achieved impressive performance – on par with the boosting-based detectors such as (Jones and Viola, 2003).

Schneiderman and Kanade (2004) described an object detector based on detecting localized parts of the object. Each part is a group of pixels or transform variables that are statistically dependent, and between parts it is assumed to be statistically independent. AdaBoost was used to compute each part's likelihood of belonging to the detected object. The final decision was made by multiplying the likelihood ratios of all the parts together and testing the result against a predefined threshold. In a later work, Schneiderman (2004a) further examined the cases where the statistical dependency cannot be easily decomposed into separate parts. He proposed a method to learn the dependency structure of a Bayesian-network-based classifier. Although the problem is known to be NP complete, he presented a scheme that selects a structure by seeking to optimize a sequence of two cost functions: the local modeling error using the likelihood ratio test as before, and the global empirical classification error computed on a cross-validation set of images. The commercial PittPatt face detection software that combines the above approach with the feature-centric cascade detection scheme in (Schneiderman, 2004b) showed state-of-the-art performance on public evaluation tests (Nechyba et al., 2007).

Schneiderman and Kanade (2004) used wavelet variables to represent parts of the faces, which do not necessarily corresponds to semantic components. In the literature, there had been many component-based object detectors that relied on semantically meaningful component detectors (Yang et al., 2002; Mohan et al., 2001; Bileschi and Heisele, 2002). In the recent work by Heisele et al. (2007), the authors used 100 textured 3D head models to train 14 component detectors. These components were initialized by a set of reference points manually annotated for the head models, and their rectangles were adaptively expanded during training to ensure good performance. The final decision was made by a linear SVM that combines all the output from the component detectors. Another closely related approach is to detect faces/humans by integrating a set of individual detectors that may have overlaps with each other. For instance, Mikolajczyk et al. (2004) applied 7 detectors to find body parts including frontal and profile faces, frontal and profile heads, frontal and profile upper body, and legs. A joint likelihood body model is then adopted to build a body structure by starting with one part and adding the confidence provided by other body part detectors.

1.4   BOOK OVERVIEW

This book will present a number of techniques related to boosting-based face detection and adaptation. More focus will be given to boosting learning algorithms instead of feature extraction. In fact, other than Chapter 5, we present results mostly with Haar-like features, which are very efficient to compute and have been shown to work very well for face detection tasks. Needless to say, the presented algorithms are generic and work well with other type of features. Below is a road map to the rest of the lecture.

CHAPTER 2–CASCADE-BASED REAL-TIME FACE DETECTION

In this chapter, we present a boosting learning framework that is capable to handle billions of training examples. It differs from traditional bootstrapping schemes in that no intermediate thresholds need to be set during training, yet the total number of negative examples used for feature selection remains constant and focused (on the poor performing ones). A multiple instance pruning scheme (Zhang and Viola, 2007) is then described to set the intermediate thresholds after boosting learning.

CHAPTER 3–MULTIPLE INSTANCE LEARNING FOR FACE DETECTION

This chapter presents two multiple instance learning schemes for face detection. The first is MilBoost (Viola et al., 2005), which is a novel boosting learning algorithm naturally integrated with multiple instance learning. MilBoost addresses the issue that it is often difficult to have different people agree upon the exact same strategy when labeling positions/scales for objects such as human faces or upper bodies. It is thus beneficial to let the algorithm figure out the objects' position and scale automatically. MilBoost achieves this by organizing positive training examples into bags and automatically electing at least one example in each bag as the true positive example during boosting learning.

The second scheme is called winner-take-all multiple category boosting (WTA-McBoost). It attempts to solve challenges faced in multiview face detection where the face pose label may be erroneous. Compared with previous approaches such as (Kim and Cipolla, 2008; Babenko et al., 2008), WTA-McBoost is derived under a different framework, and it is much more efficient for training. Feature sharing (Torralba et al., 2004) can also be easily integrated, making the detector very efficient to run during testing as well.

CHAPTER 4–DETECTOR ADAPTATION

In many applications, a detector trained with generic data sets may not perform optimally in a new environment. Detection adaption is a promising solution for this problem. We present an adaptation scheme based on the Taylor expansion of the boosting learning target function, and we propose to store the second order statistics of the generic training data for future adaptation (Zhang et al., 2008a). We will show that with a small amount of labeled data in the new environment, the detector's performance can be greatly improved.

CHAPTER 5–OTHER APPLICATIONS

We present two interesting applications where boosting learning was applied successfully. The first application is face verification for filtering and ranking image/video search results on celebrities. We present boosted multi-task learning (MTL) (Wang et al., 2009b), yet another boosting learning algorithm that extends MilBoost with a graphical model. Since the available number of training images for each celebrity may be limited, learning individual classifiers for each person may cause overfitting. MTL jointly learns classifiers for multiple people by sharing a few boosting classifiers in order to avoid overfitting. The effectiveness of boosted MTL is shown with a large number of celebrity images/videos from the web.

The second application deals with the problem of speaker detection in conference rooms (Zhang et al., 2008b). The goal is to find who is speaking in a conference room, given a microphone array and a panoramic video of the room. We show that by combining audio and visual features in a boosting framework, we can determine the speaker's position very accurately. The algorithm has been integrated with the RoundTable device and shipped to thousands of users in the summer of 2007.

CHAPTER 6–CONCLUSIONS AND FUTURE WORK

Conclusions and future work will be given at the end of this lecture.

CHAPTER 2

Cascade-based Real-Time Face Detection

Learning a cascade-based face detector used to be a very time-consuming task. In particular, early approaches such as (Viola and Jones, 2001; Lienhart and Maydt, 2002) used manual tuning or heuristics to set the intermediate rejection thresholds for the detector, which is inefficient and suboptimal. Recently, various approaches has been proposed to address this issue. Notably, Bourdev and Brandt (Bourdev and Brandt, 2005) proposed a method for setting rejection thresholds based on an ad hoc detection rate target called a "rejection distribution vector", which is a parameterized exponential curve. Like the original Viola-Jones proposal, the soft-cascade gradually gives up on a number of positive examples in an effort to aggressively reduce the number of negatives passing through the cascade. Perhaps a particular family of curves is more palatable, but it is still arbitrary and non-optimal. Sochman-Matas (Sochman and Matas, 2005) used a ratio test to determine the rejection thresholds. While this has statistical validity, distributions must be estimated, which introduces empirical risk. This is a particular problem for the first few rejection thresholds, and it can lead to low detection rates on test data.

In Section 2.3 of this chapter, we introduce a new mechanism for setting the rejection thresholds of any soft-cascade which is conceptually simple, has no tunable parameters beyond the final detection rate target, yet yields a cascade which is both highly accurate and very fast. Training data is used to set all reject thresholds after the final classifier is learned. There are no assumptions about probability distributions, statistical independence, or ad hoc intermediate targets for detection rate (or false positive rate). The approach is based on two key insights that constitute the major contributions of this chapter: 1) positive examples that are rejected by the complete classifier can be safely rejected earlier during pruning, 2) each ground-truth face requires no more than one matched detection window to maintain the classifier's detection rate. We propose a novel algorithm, multiple instance pruning (MIP), to set the reject thresholds automatically, which results in a very efficient cascade detector with superior performance.

Regarding how the soft-cascade shall be trained, in Section 2.1, we present a statistically sound approach that uses importance sampling and weight trimming to learn from billions of negative examples. Weight trimming can dramatically reduce computation for boosted methods without sacrificing accuracy. In our approach, no example is discarded permanently; therefore, it is robust to accidental drops of negative examples during learning, which was a key concern for setting hard thresholds during training.

Another useful technique that will be introduced in Section 2.2 of this chapter is the so-called "fat stumps", which is a histogram-based weak classifier that works on a single feature. Similar to the Bayesian stump in (Xiao et al., 2007), the thresholds of fat stumps are derived from iterative split and merge operations instead of being at equal distances and fixed. Experiments show that fat stumps are very effective in reducing the total number of features used in a detector, and they result in high performance detectors.

2.1   SOFT-CASCADE TRAINING

Face detection is a rare event detection task. Therefore, training a face detector often involves billions of negative training examples. To train a detector with so many negative examples the traditional approach has been through bootstrapping (Viola and Jones, 2001; Li et al., 2002; Xiao et al., 2003). That is, a node classifier is first trained with a subset of the negative examples. A threshold is then set for the classifier, which is subsequently used to scan the remaining negative examples to find those that have not been rejected. These negative examples are then added to the subset for learning the next node classifier. With bootstrapping, the classifiers' thresholds have to be set during training, which is the root cause of many inefficiencies. If the thresholds are set too aggressively, the final detector will be fast, but the detection rate may drop. On the other hand, if the thresholds are set conservatively, many negative examples will pass the early nodes, making both the training and testing procedure slow.

The solution is thus to eliminate threshold setting during training, and set them afterwards (Luo, 2005; Bourdev and Brandt, 2005). Nevertheless, due to the large amount of data, training a soft-cascade without setting any intermediate thresholds is a non-trivial task. Bourdev and Brandt (2005) proposed a scheme that starts with a small set of training examples and grows the training set gradually. The number of new misclassified non-faces to be added at each training cycle affects the focus of AdaBoost during training. If the number is too large, AdaBoost may not be able to catch up and the false positive rate will be high. If the number is too small, the cascade may contain too many weak classifiers in order to reach a reasonable false positive rate. Since the training set always grows, later stages of the training will be slower and slower. In the following, we present an importance-sampling-based scheme that combines both weight trimming and bootstrapping to solve the problem in a principled way.

The flowchart of our training procedure is shown in Fig. 2.1. All the training examples are initially stored on a disk drive. Denote the training example set as $S = \{(x_i, z_i), i = 1, \cdots, N\}$, where $z_i = 0$ for negative examples and $z_i = 1$ for positive examples. Let N_+ be the number of positive examples and N_- be the number of negative examples. $N_+ + N_- = N$, and N_- is usually on the order of billions. Each training example are associated with an AdaBoost weight. Recall the target criterion for AdaBoost is:

$$L^T = \sum_{i=1}^{N} \left[I(z_i = 1) \exp\{-F^T(x_i)\} + I(z_i = 0) \exp\{F^T(x_i)\} \right], \tag{2.1}$$

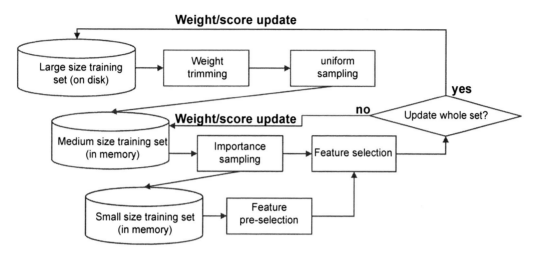

Figure 2.1: The flowchart of the proposed training procedure using importance sampling and weight trimming.

where T is the number of weak classifiers so far, and $I(\cdot)$ is the indicator function. For an example instance x_i, the score up to weak classifier T is $F^T(x_i)$, and the corresponding weight is $w_i = \exp\{-F^T(x_i)\}$ for positive examples and $w_i = \exp\{F^T(x_i)\}$ for negative examples.

The first step in the flowchart is weight trimming (Friedman et al., 1998). The *negative* training examples with the smallest weights are trimmed in this stage, up to a percentage of the total weight, say, between 1% and 10%. Since the weights are typically very skewed toward a small number of hard examples, this can eliminate a very large portion of the examples. It was shown in (Friedman et al., 1998) that weight trimming can dramatically reduce computation for boosted methods without sacrificing accuracy. A nice property of weight trimming is that examples are never thrown away completely because even if a certain example is trimmed at one stage, it may return at a later stage. Since no example is discarded permanently, it is ideal for learning a soft cascade.

After weight trimming, we uniformly sample Q negative examples from the remaining ones. Combined with all the positive training examples, they form a medium size training set which will be kept in memory. Q is usually on the order of several millions, determined by the available computer memory.

The medium size training set is still too large for direct feature selection, as typically there are over one hundred thousand Haar features available for a patch size of 24×24 pixels in our training. We overcome the issue by another round of sampling. This time, the sampling is performed based on the example weights. Namely, examples with higher weights are more likely to be sampled, while examples with lower weights are less likely to be sampled. The result of such an importance sampling step is a small size training set, typically a few thousand examples, on which all the Haar feature values can be computed and stored in memory. The feature pre-selection step can thus be performed efficiently, resulting in K top features. The typical value of K is 20.

We then conduct feature selection on the medium size training set, though only focusing on the K pre-selected top features. Once the best feature is selected, the weights and scores of the medium size training set will be updated accordingly. After a few rounds of feature selection, the training program updates the whole training set on the disk, and it goes through another round of weight trimming and uniform sampling to obtain new examples from the large data set. The pseudo code of the training procedure is summarized in Fig. 2.2.

One issue for the importance-sampling-based scheme when applied to the original AdaBoost/RealBoost framework is the rapid increase of weights for some examples due to the exponential criterion. After a few rounds of feature selection, the weights will be skewed toward a very small number of training examples. These examples are either hard examples or outliers. Consequently, the importance sampling step will likely only pick very few examples, leading to ineffective feature selection. We instead adopt the logistic variant of AdaBoost developed by Collins et al. (2002) for training. Instead of using the exponential function to compute the examples' weights, we use the logistic function:

$$
w_i = \begin{cases} \dfrac{1}{1+\exp\{F^T(x_i)\}}, & z_i = 1; \\[2ex] \dfrac{1}{1+\exp\{-F^T(x_i)\}}, & z_i = 0. \end{cases} \tag{2.2}
$$

Since the logistic function is bounded between 0 and 1 when $F^T(x_i)$ goes to $\pm\infty$, it is more friendly to importance sampling. Our target criterion for boosting learning is now:

$$
L^T = \sum_{i=1}^{N}\left[I(z_i = 1)\frac{1}{1 + \exp\{F^T(x_i)\}} + I(z_i = 0)\frac{1}{1 + \exp\{-F^T(x_i)\}}\right]. \tag{2.3}
$$

This is indeed the target criterion for LogitBoost (Friedman et al., 1998). The feature selection criterion described in Section 1.2.2, in particular, the Z score (Eq. (1.12)) will not be the optimal feature selection criterion any more. Friedman et al. (1998) presented an adaptive Newton steps method for LogitBoost training, but it can be slow for large data set training. Note, for positive examples, the scores $F^T(x_i)$ tend to be positive, and $\frac{1}{1+\exp\{F^T(x_i)\}} \approx \exp\{-F^T(x_i)\}$, when $F^T(x_i) \to +\infty$. Similarly, for negative examples, the scores $F^T(x_i)$ tend to be negative, and $\frac{1}{1+\exp\{-F^T(x_i)\}} \approx \exp\{F^T(x_i)\}$, when $F^T(x_i) \to -\infty$. The Z score is thus still a good criterion for feature selection. Another important benefit is that the Z score can be computed very efficiently for exhaustive feature selection; hence, we still use it exclusively in our experimental results despite the change in how the weights are computed.

Since data labeling is a very expensive task, the ratio between positive and negative examples in the training set may be very different from the test environment. Instead of introducing asym-

Input

- Training set $S = \{(x_i, z_i), i = 1, \cdots, N\}$, where $z_i = 0, 1$ for negative and positive examples.
- A ratio r between the number of positive an negative examples for future testing sets.
- T is the total number of weak classifiers, which can be set through cross-validation.

Initialize

- Take all positive examples and randomly sample Q negative examples from the training set.
- Initialize negative example weight factor $\eta_{neg} = \frac{N_+}{rQ}$. All sampled training examples are given an initial score $s_{init} = \ln r$.

Adaboost Learning

For $t = 1, \cdots, T$:

1. Compute weights for the medium size training set. For positive examples, $w_i^t = \frac{1}{1+\exp\{F^T(x_i)\}}$, for negative examples, $w_i^t = \frac{\eta_{neg}}{1+\exp\{-F^T(x_i)\}}$.
2. Use importance sampling to generate the small size training set, conduct feature pre-selection using fat stumps (Section 2.2), Bayesian stumps (Xiao et al., 2007) or LUT classifiers (Wu et al., 2004a).
3. Select the best feature from the top features chosen during the previous feature pre-selection step, and construct the weak classifier f_t with the associated confidences.
4. Update scores of all examples in the medium size training set.
5. If t belong to set $A = \{2, 4, 8, 16, 32, \cdots\}$,

   - Update scores and weights of the whole training set using the previously selected weak classifiers $f_1, \cdots, f_t$.
   - Perform weight trimming to trim 10% of the negative weights.
   - Take all positive examples and randomly sample Q negative examples from the trimmed training set.
   - Update negative example weight ratio η_{neg} (Eq. (2.6)).

6. Set preliminary rejection threshold $\theta(t)$ of $\sum_{j=1}^{t} \alpha_j f_j$ as the minimum score of all positive examples at stage t.

Output

Series of weak classifiers $f_t, t = 1, \cdots, T$ and the associated confidences, and the preliminary rejection thresholds $\theta(t)$.

Figure 2.2: Adaboost learning with weight trimming and boostrapping.

metric boosting losses (Viola and Jones, 2002; Wu et al., 2005; Masnadi-Shirazi and Vasconcelos, 2007a), which did not demonstrate superior performance over regular boosting (Wu et al., 2005), we introduce a ratio r between the number of positive and negative examples that we expect to see in future testing sets. This ratio is maintained as a negative example weight factor η_{neg} throughout the training process, as shown in Fig. 2.2.

During initialization, since we have a total of N_+ positive examples, we need the Q sampled negative examples to be equivalent to rN_+ negative examples. Hence, the initial weight factor for negative examples is:

$$\eta_{\text{neg}} = \frac{N_+}{rQ}.$$ (2.4)

The initial score s_{init} of all the examples in the sampled set shall be the same. To ensure that the total weight of positive examples and the total weight of negative examples are equal, s_{init} needs to satisfy:

$$\frac{r}{1 + e^{s_{\text{init}}}} = \frac{1}{1 + e^{-s_{\text{init}}}}.$$ (2.5)

Consequently, $s_{\text{init}} = \ln r$.

During the AdaBoost learning stage, the medium size training set uniformly sampled from the original training set is used for feature selection. Note the negative examples' weights are multiplied by the factor η_{neg} in order to approximate the correct example ratio between positive and negative examples. From time to time, when $t \in A$, where A is a user-defined set, the training process will update the scores and weights of the whole training set on the disk. One can, in theory, update the scores of the whole training set after each feature is learned if computationally affordable, though the gain in detector performance may not be visible. In addition, the negative example weight factor will be updated in this step. Based on $w_i^t = \frac{1}{1+\exp\{-F^t(x_i)\}}$, let the total weight of the N_- negative examples be $W_{N_-}^t$, and that of the resampled Q negative examples be W_Q^t, we have:

$$\eta_{\text{neg}} = \frac{N_+ W_{N_-}^t}{r N_- W_Q^t}.$$ (2.6)

Note, a set of thresholds are also returned by this process. These preliminary rejection thresholds are extremely conservative, retaining *all* positive examples in the training set. They result in a very slow detector – the average number of features visited per window is on the order of hundreds. These thresholds will be replaced with the ones derived by the multiple instance pruning algorithm in Section 2.3. We set the preliminary thresholds only to moderately speed up the computation of ROC curves before MIP.

2.2   FAT STUMPS

We replace the simple decision stumps used by Viola and Jones (2001) with "fat stumps", a decision tree which performs a multi-way split based on multiple thresholds, as shown in Fig. 2.3. A confidence

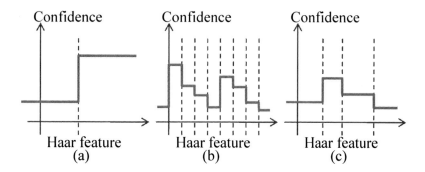

Figure 2.3: Various weak classifiers – (a) Stump classifier, (b) LUT classifier and (c) Fat stump multi-threshold classifier.

or vote is given to the examples falling into each interval of the Haar feature. Multiple thresholds can allow more flexibility and help extract more information from a single rectangular filter output. For example, a filter response that indicates a face if its value is either very large or very small. The motivation is similar to that of a look up table (Wu et al., 2004a), but fat stumps have the advantage of dynamically determined bin widths and positions.

In the literature, Brubaker et al. (2005) showed that using full CART decision trees as weak classifiers leads to an improvement in detection performance. This comes with a measurable increase in detection time. Wu et al. (2004a) proposed to use a look up table (LUT) with fixed bins instead of simple stumps. However, the performance of the final detector can be sensitive to the number of bins for the LUT. Too fine granularity of the histogram may cause overfitting, while too coarse granularity may lead to poor performance due to the fixed boundaries between bins. In contrast, the fat stumps classifiers split the input range into a small number of flexible, rather than fixed, bins. They are less likely to overfit, yet can provide significant improvement over traditional decision stumps.

The RealBoost criterion for feature selection with multiple thresholds has been given in (Wu et al., 2004a). It is a straightforward extension to the single threshold case presented in Section 1.2.2. In particular, assume a set of decision thresholds $H_1, \cdots, H_{J-1}$ divide the output of feature $h(x)$ into J subregions, $u_1, \cdots, u_J, u_1 \cup \cdots \cup u_J = \mathbb{R}$. The base function $f(x)$ is thus:

$$f(x) = c_j, \text{ if } h(x) \in u_j \tag{2.7}$$

where $c_1, \cdots, c_J$ are confidences. Let

$$
\begin{aligned}
W_{+j} &= \sum_i I(f(x_i) \in u_j) I(z_i = 1) \exp\{-F^t(x_i)\} \\
W_{-j} &= \sum_i I(f(x_i) \in u_j) I(z_i = 0) \exp\{F^t(x_i)\}.
\end{aligned}
\tag{2.8}
$$

The target criterion can thus be written as:

$$L^{t+1} = \sum_{j=1}^{J} \left[W_{+j} e^{-c_j} + W_{-j} e^{c_j} \right].$$ (2.9)

We can see L^{t+1} is minimized when

$$c_j = \frac{1}{2} \ln \left(\frac{W_{+j}}{W_{-j}} \right).$$ (2.10)

Plugging into Eq. (2.9), we have:

$$L^{t+1} = 2 \sum_{j=1}^{J} \sqrt{W_{+j} W_{-j}}.$$ (2.11)

We still refer Eq. (2.11) as the Z score. At iteration $t + 1$, for every Haar-like feature $h(x)$, we need to find the optimal threshold set $H_1, \cdots, H_{j-1}$ to minimize the Z score L^{t+1}. We propose to solve the problem in an iterative fashion, as described in Fig. 2.4. Note convergence is guaranteed because at each iteration the Z score does not increase. However, there is no guarantee that it will converge to the global minimum of the Z score. In our current implementation, the maximum number of iterations is set as 5. We find 1-3 iterations is usually enough for convergence. Feature selection is thus still reasonably fast to conduct, even with 7-15 dynamic thresholds.

The proposed weak classifier can be considered equivalent to a small decision tree or CART, but each node in the tree uses the same rectangle filter. This improves the speed of each classifier over CART while still achieving superior performance.

2.3 MULTIPLE INSTANCE PRUNING

The soft cascade detector trained in Section 2.1 contains only a set of very conservative intermediate thresholds. In this section, we discuss how to prune the detector in order to improve its detection speed.

2.3.1 PRUNING USING THE FINAL CLASSIFICATION

We start by proposing a scheme which is simultaneously simpler and more effective than earlier techniques. Our key insight is quite simple: the reject thresholds are set so that they give up on precisely those positive examples that are rejected by the complete classifier. Note that the score of each example, $F^t(x_i)$, can be considered as a trajectory through time. The full classifier rejects a positive example if its final score $F^T(x_i)$ falls below the final threshold $\theta(T)$. In the simplest version of our threshold setting algorithm, all trajectories from positive windows which fall below the final threshold are removed. The intermediate rejection thresholds are then simply:

$$\theta(t) = \min_{\left\{ i \middle| F^T(x_i) > \theta(T), z_i = 1 \right\}} F^t(x_i)$$ (2.12)

Input

- Histograms of weights of positive and negative examples for a certain feature.
- Number of thresholds to be used J.

Initialize

- Find a single threshold to minimize the Z score (a stump classifier).
- Repeatedly split the segment with the highest partial Z score $\sqrt{W_{+j}W_{-j}}$ using the same algorithm of finding a single threshold.

Iterative Threshold Refinement

Iterate till convergence or maximum number of iterations reached:

For $j = 1, \cdots, J - 1$,

- Fix all other thresholds, adjust threshold H_j locally to minimize the overall Z score.

Output

Set of feature thresholds H_j, $j = 1, \cdots, J - 1$ and confidence in each region $c^j = \frac{1}{2} \ln\left(\frac{W_{+j}}{W_{-j}}\right)$, $j = 1, \cdots, J$.

Figure 2.4: Algorithm for learning a fat stump weak classifier.

where $\{x_i, z_i\}$ is the training set in which $z_i = 1$ indicates positive windows and $z_i = 0$ indicates negative windows. These thresholds produce a reasonably fast classifier which is guaranteed to produce no more errors than the complete classifier on the training data set. We call this pruning algorithm *direct backward pruning* (DBP).

One might question whether the minimum of all retained trajectories is robust to mislabeled or noisy examples in the training set. Note that the final threshold of the complete classifier will often reject mislabeled or noisy examples (though they will be considered false negatives). These rejected examples play no role in setting the rejection thresholds. We have found this procedure very robust to the types of noise present in real training sets.

In past approaches, thresholds are set to reject the largest number of negative examples and only a small *percentage of positive examples*. These approaches justify these thresholds in different ways, but they all struggle to determine the correct percentage accurately and effectively. In the new approach, the final threshold of the *complete soft-cascade* is set to achieve the require detection rate. Rejection thresholds are then set to reject the largest number of negative examples and retain *all* positive examples that are retained by the complete classifier. The important difference is that the particular positive examples which are rejected are those which are destined to be rejected by the final classifier. This yields a fast classifier which *labels all positive examples in exactly the same way as*

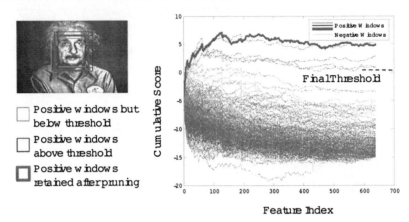

Figure 2.5: Traces of cumulative scores of different windows in an image of a face. See text.

the complete classifier. In fact, it yields the fastest possible soft-cascade with such property (provided that the weak classifiers are not re-ordered). Note, some negative examples that eventually pass the complete classifier threshold may be pruned by earlier rejection thresholds. This has the satisfactory side benefit of reducing false positive rate as well. In contrast, although the detection rate on the training set can also be guaranteed in the algorithm by Bourdev and Brandt (2005), there is no guarantee that false positive rate will not increase.

Bourdev and Brandt proposed to reorder the weak classifiers based on the separation between the mean score of the positive examples and the mean score of the negative examples. Our approach is equally applicable to a reordered soft-cascade.

Fig. 2.5 shows 293 trajectories from a single image whose final score is above −15. While the rejection thresholds are learned using a large set of training examples, this one image demonstrates the basic concepts. The red trajectories are negative windows. The single physical face is consistent with a set of positive detection windows that are within an acceptable range of positions and scales. Typically, there are tens of acceptable windows for each face. The blue and magenta trajectories correspond to acceptable windows that fall above the final detection threshold. The cyan trajectories are potentially positive windows that fall below the final threshold. Since the cyan trajectories are rejected by the final classifier, rejection thresholds need only retain the blue and magenta trajectories.

In a sense, the complete classifier, along with a threshold which sets the operating point, provides labels on examples that are *more valuable* than the ground-truth labels. There will always be a set of "positive" examples that are extremely difficult to detect, or worse that are mistakenly labeled positive. In practice, the final threshold of the complete classifier will be set so that these *particular examples* are rejected. In our new approach, these particular examples can be rejected early in the computation of the cascade. Compared with existing approaches, which set the reject thresholds in a heuristic manner, our approach is data-driven and hence more principled.

2.3.2   MULTIPLE INSTANCE PRUNING

The notion of an "acceptable detection window" plays a critical role in an improved process for setting rejection thresholds. It is difficult to define the correct position and scale of a face in an image. For a purely upright and frontal face, one might propose the smallest rectangle which includes the chin, forehead, and the inner edges of the ears. But, as we include a range of non-upright and non-frontal faces these rectangles can vary quite a bit. Should the correct window be a function of apparent head size? Or is eye position and interocular distance more reliable? Even given clear instructions, one finds that two subjects will differ significantly in their "ground-truth" labels.

Recall that the detection process scans the image generating a large, but finite, collection of overlapping windows at various scales and locations. Even in the absence of ambiguity, some flexibility is required to ensure that at least one of the generated windows is considered a successful detection for each face. Experiments typically declare that any window which is within 50% in size and within a distance of 50% (of size) be considered a true positive. Using typical scanning parameters this can lead to tens of windows which are all equally valid positive detections. If any of these windows is classified positive then this face is consider detected. Therefore, one need only retain one "acceptable" window for each face that is detected by the final classifier. A more aggressive threshold is thus defined as:

$$\theta(t) = \min_{k \in P} \left[\max_{\left\{ i \,\middle|\, i \in F_k \cap R_k, z_i = 1 \right\}} F^t(x_i) \right] \quad (2.13)$$

where k is the index of ground-truth faces; F_k is the set of acceptable windows associated with ground-truth face k and R_k is the set of windows which are "retained" (see below). P is the set of ground-truth faces that have at least one acceptable window above the final threshold:

$$P = \left\{ k \,\middle|\, \max_{\left\{ i \,\middle|\, i \in F_k \right\}} F^T(x_i) > \theta(T) \right\} \quad (2.14)$$

In this new procedure, the acceptable windows come in bags, only one of which must be classified positive in order to ensure that each face is successfully detected. This new criteria for success is more flexible and therefore more aggressive. We call this pruning method *multiple instance pruning* (MIP).

Returning to Fig. 2.5, we can see that the blue, cyan, and magenta trajectories actually form a "bag". Both in this algorithm and in the simpler DBP algorithm, the cyan trajectories are rejected before the computation of the thresholds. The benefit of this new algorithm is that the blue trajectories can be rejected as well.

The definition of "retained" examples in the computation above is a bit more complex than before. Initially, the trajectories from the positive bags that fall above the final threshold are retained. The set of retained examples is further reduced as the earlier thresholds are set. This is in contrast to the DBP algorithm where the thresholds are set to preserve *all* retained positive examples. In the new algorithm, the partial score of an example can fall below the current threshold (because it is in

Input

- A cascade detector.
- Threshold $\theta(T)$ at the final stage of the detector.
- A large training set (the whole training set to learn the cascade detector can be reused here).

Initialize

- Run the detector on all rectangles that match with any ground-truth faces. Collect all windows that are above the final threshold $\theta(T)$. Record all intermediate scores as $s(k, j, t)$, where $k = 1, \cdots, K$ is the face index; $j = 1, \cdots, J_k$ is the index of windows that match with face k; $t = 1, \cdots, T$ is the index of the feature node.
- Initialize flags $f(k, j)$ as true.

MIP
For $t = 1, \cdots, T$:

1. For $k = 1, \cdots, K$: find $\hat{s}(k, t) = \max_{\{j \mid f(k,j)=\text{true}\}} s(k, j, t)$.
2. Set $\theta(t) = \min_k \hat{s}(k, t) - \epsilon$ as the rejection threshold of node t, $\epsilon = 10^{-6}$.
3. For $k = 1, \cdots, K, j = 1, \cdots, J_k$: set $f(k, j)$ as false if $s(k, j, t) < \theta(t)$.

Output
Rejection thresholds $\theta(t), t = 1, \cdots, T$.

Figure 2.6: The MIP algorithm.

a bag with a better example). Each such example is removed from the retained set R_k and not used to set subsequent thresholds.

The pseudo code of the MIP algorithm is shown in Fig. 2.6. It guarantees the same face detection rate on the training dataset as the complete classifier. Note that the algorithm is greedy, setting earlier thresholds first so that all positive bags are retained and the fewest number of negative examples pass. Theoretically, it is possible that delaying the rejection of a particular example may result in a better threshold at a later stage. Searching for the optimal MIP pruned detector, however, may be quite expensive. The MIP algorithm is *guaranteed to generate a soft-cascade that is at least as fast as DBP* since the criteria for setting the thresholds is less restrictive.

2.4   EXPERIMENTAL RESULTS

More than 20,000 images were collected from the web, containing roughly 10,000 faces. Over 2 billion negative examples are generated from the same image set. A soft cascade classifier is learned through the learning algorithm presented in Fig. 2.2.

The training process was conducted on a dual core AMD Opteron 2.2 GHz processor with 16 GB of RAM. It takes less than 2 days to train a classifier with 700 fat stumps weak classifiers based on the Haar features, in contrast to months for the first Viola-Jones detector (Viola and Jones, 2001). The testing set is the standard MIT+CMU frontal face database (Sung and Poggio, 1998; Rowley et al., 1998), which consists of 125 grayscale images containing 483 labeled frontal faces. A detected rectangle is considered to be a true detection if the distance from its center to the ground truth center is less than 50% of the width of the ground truth rectangle, and the difference in scale is also within 50% of the ground truth scale.

It is difficult to compare the performance of various detectors since every detector is trained on a different dataset. Nevertheless, we show the ROC curves of a number of existing detectors and ours in Fig. 2.7(a). Note there are two curves plotted for soft cascade. The first curve has very good performance at the cost of slow speed (average 37.1 features per window). The classification accuracy dropped significantly in the second curve, which is faster (average 25 features per window).

Fig. 2.8(a) compares DBP and MIP with different final thresholds of the strong classifier. The original data set for learning the soft cascade is reused for pruning the detector. Since MIP is a more aggressive pruning method, the average number of features evaluated is much lower than DBP. Note both DBP and MIP guarantee that no positive example from the *training set* is lost. There is no similar guarantee for test data, though.

Fig. 2.7(b) shows that there is no practical loss in classification accuracy on the MIT+CMU test dataset for various applications of the MIP algorithm (note that the MIT+CMU data is not used by the training process in any way).

Speed comparison with other algorithms are subtle (Fig. 2.8(b)). The first observation is that higher detection rates almost always require the evaluation of additional features. This is certainly true in our experiments, but it is also true in past papers (e.g., the two curves of Bourdev-Brandt soft cascade in Fig. 2.7(a)). The fastest algorithms often cannot achieve very high detection rates. One explanation is that in order to achieve higher detection rates one must retain windows that are "ambiguous" and may contain faces. The proposed MIP-based detector yields a much lower false positive rate than the 25-feature Bourdev-Brandt soft cascade and nearly 35% improvement on detection speed. While the WaldBoost algorithm is quite fast, detection rates are measurably lower. Detectors such as Viola-Jones, boosting chain, FloatBoost, and Wu et al. all requires manual tuning. We can only guess how much trial and error went into getting a fast detector that yields good results.

The expected computation time of the DBP soft-cascade varies monotonically in detection rate. This is guaranteed by the algorithm. In experiments with MIP, we found a surprising quirk in the expected computation times. One would expect that if the required detection rate is higher, it world be more difficult to prune. In MIP, when the detection rate increases, there are two conflicting factors involved. First, the number of detected faces increases, which increases the difficulty of pruning. Second, for each face, the number of retained and acceptable windows increases. Since we are computing the maximum of this larger set, MIP can in some cases be more aggressive. The second factor explains the increase of speed when the final threshold changes from -1.5 to -2.0.

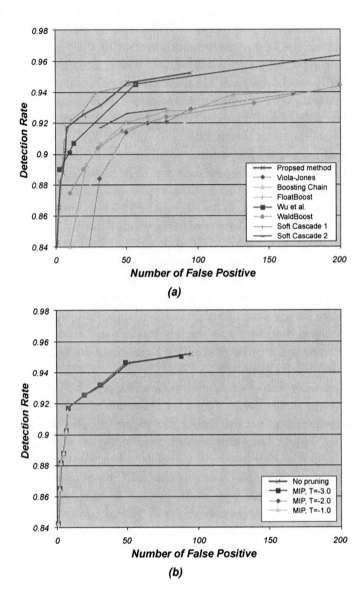

Figure 2.7: (a) Performance comparison with existing works (e.g., Viola-Jones (Viola and Jones, 2001), boosting chain (Xiao et al., 2003), FloatBoost (Li et al., 2002), Wu et al. (Wu et al., 2004a), Wald-Boost (Sochman and Matas, 2005) and soft cascade (Bourdev and Brandt, 2005)) on MIT+CMU frontal face dataset. (b) ROC curves of the detector after MIP pruning using the original training set. No performance degradation is found on the MIT+CMU testing dataset.

Final Threshold	-3.0	-2.5	-2.0	-1.5	-1.0	-0.5	0.0
Detection Rate	95.2%	94.6%	93.2%	92.5%	91.7%	90.3%	88.8%
# of False Positive	95	51	32	20	8	7	5
DBP	36.13	35.78	35.76	34.93	29.22	28.91	26.72
MIP	16.11	16.06	16.80	18.60	16.96	15.53	14.59

(a)

Approach	Viola-Jones	Boosting chain	FloatBoost	WaldBoost	Wu et al.	Soft cascade
Total # of features	6061	700	2546	600	756	4943
Slowness	10	18.1	18.9	13.9	NA	37.1 (25)

(b)

Figure 2.8: (a) Pruning performance of DBP and MIP. The bottom two rows indicate the average number of features visited per window on the MIT+CMU dataset. (b) Results of existing work. Slowness is again measured by the average number of features visited per window.

The direct performance comparison between MIP and Bourdev-Brandt (B-B) was performed using the same soft-cascade and the same data. In order to better measure performance differences, we created a larger test set containing 3,859 images with 3,652 faces collected from the web. Both algorithms prune the strong classifier for a target detection rate of 97.2% on the *training set*, which corresponds to having a final threshold of -2.5 in Fig. 2.8(a). We use the same exponential function family as Bourdev and Brandt (2005) for B-B, and we adjust the control parameter α in the range between -16 and 4. The results are shown in Fig. 2.9. It can be seen that the MIP pruned detector has the best detection performance. When a positive α is used (e.g., $\alpha = 4$), the B-B pruned detector is still worse than the MIP pruned detector, and its speed is 5 times slower (56.83 vs. 11.25). On the other hand, when α is negative, the speed of B-B pruned detectors improves and can be faster than MIP (e.g., when $\alpha = -16$). Note, adjusting α leads to changes both in detection time and false positive rate.

In practice, both MIP and B-B can be useful. MIP is fully automated and guarantees detection rate with no increase in false positive rate on the training set. The MIP pruned strong classifier is usually fast enough for most real-world applications. On the other hand, if speed is the dominant factor, one can specify a target detection rate and target execution time and use B-B to find a solution. Note such a solution is not guaranteed, and the false positive rate may be unacceptably high (The performance degradation of B-B heavily depends on the given soft-cascade. While with our detector the performance of B-B is acceptable even when $\alpha = -16$, the performance of the detector in (Bourdev and Brandt, 2005) drops significantly from 37 features to 25 features, as shown in Fig. 2.7 (a)).

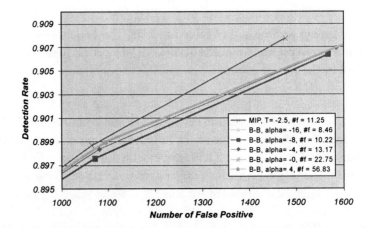

Figure 2.9: The detector performance comparison after applying MIP and Bourdev-Brandt's method Bourdev and Brandt (2005). Note, this test was done using a much larger, and more difficult, test set than MIT+CMU. In the legend, symbol #f represents the average number of weak classifiers visited per window.

CHAPTER 3

Multiple Instance Learning for Face Detection

The multiple instance pruning algorithm introduced in the previous section utilized the fact that for a ground truth face in an image, numerous detection results surrounding the ground truth rectangle are plausible. Such an observation is indeed quite general. In many object recognition tasks, it is often extremely tedious to generate large training sets of objects because it is not easy to specify exactly where the objects are. For instance, given a ZIP code of handwritten digits, which pixel is the location of a "5"? This sort of ambiguity leads to training sets which themselves have high error rates, which limits the accuracy of any trained classifier.

In this chapter, we explicitly acknowledge that object recognition is innately a multiple instance learning (MIL) problem: we know that objects are located in regions of the image, but we don't know exactly where. In MIL, training examples are not singletons. Instead, they come in "bags", where all of the examples in a bag share a single label (Dietterich et al., 1997). A positive bag means that at least one example in the bag is positive, while a negative bag means that all examples in the bag are negative. In MIL, learning must simultaneously learn which examples in the positive bags are positive, along with the parameters of the classifier. We present a boosting-based MIL algorithm called MILBoost in Section 3.1. MILBoost is based on the AnyBoost framework by Mason et al. (2000), which views boosting as a gradient decent process. The derivation builds on previous appropriate MIL cost functions, namely integrated segmentation and recognition (ISR) (Keeler et al., 1990) and Noisy OR (Heckerman, 1989). Experiments conducted on low resolution face detection (upper body detection) show that MILBoost can produce classifiers with much higher detection rates than regular AdaBoost.

In Section 3.2, we further extend the multiple instance learning idea to multiple category learning problems, e.g., multiview face detection. The positive examples are divided into multiple subcategories for training subcategory classifiers individually. However, the subcategory labeling process, either through manual labeling or through clustering, is suboptimal for the overall classification task. We propose two novel multiple category boosting (McBoost) algorithms, the Probabilistic McBoost (Prob-McBoost) and the Winner-take-all McBoost (WTA-McBoost), which overcome the above issue through soft labeling. In Prob-McBoost, a set of likelihood values that describe the probability of each example belonging to each subcategory is kept during training. These likelihood values are combined to compute the probability of the example being a positive example. The learning algorithm then maximizes the overall probability of all examples in the training data set. In WTA-McBoost, each positive example has a unique subcategory label at any stage of the training process, but the label may switch to a different subcategory if a higher score is achieved by that subcategory classifier. By allowing examples to self-organize themselves in such a winner-take-all manner, WTA-McBoost outperforms traditional schemes significantly, as supported by our experiments on learning a multi-view face detector.

3.1   MILBOOST

The idea for multiple instance learning was originally proposed in 1990 for handwritten digit recognition by Keeler et al. (1990). Their approach was called Integrated Segmentation and Recognition (ISR). In that paper, the position of a digit in a ZIP code was considered completely unknown. ISR simultaneously learned the positions of the digits and the parameters of a convolutional neural network recognizer. More details on ISR will be given in Section 3.1.2.

Another relevant example of MIL is the Diverse Density approach of Maron and Lozano-Perez (1998). Diverse Density uses the Noisy OR generative model (Heckerman, 1989) to explain the bag labels. A gradient-descent algorithm was used to find the best point in input space that explains the positive bags. We also utilize the Noisy OR generative model in a version of our algorithm, as presented in Section 3.1.1.

One early work that partially inspired our approach is the work by Nowlan and Platt (1995), which built on the earlier work of Keeler et al. (1990). In that paper, a convolutional neural network was trained to detect hands. The exact location and size of the hands is approximately truthed: the neural network used MIL training to co-learn the object location and the parameters of the classifier. The system is effective, though not as fast as the typical boosting-based cascade detectors.

In the following, we will present our ISR- and Noisy-OR-based approach to solve the MIL problem. The derivation uses the AnyBoost framework of Mason et al. (2000), which views boosting as a gradient descent process. Since the Noisy OR derivation is simpler and a bit more intuitive, we describe it first.

3.1.1   NOISY-OR MILBOOST

In MILBoost, examples are not individually labeled. Instead, they reside in bags, and each bag is given a classification label. For instance, consider for a two-class classification problem, a set of labeled examples $S = \{(x_{ij}, z_i), i = 1, \cdots, N, j = 1, \cdots, J_i\}$ are given, where x_{ij} belongs to a domain or instance space $\mathcal{X}$, and $z_i = 1$ for positive examples and $z_i = 0$ for negative examples. Note each example is indexed with two indices: i, which indexes the bag, and j, which indexes the example within each bag. Recall in boosting each example is classified by a linear combination of weak classifiers. Let $y_{ij} = F^T(x_{ij}) = \sum_t \lambda^t f^t(x_{ij})$ be the weighted sum of weak classifiers, often referred as the *score* of example x_{ij}. The probability of an example x_{ij} being positive is given by:

$$p_{ij} = \frac{1}{1 + \exp(-y_{ij})}, \tag{3.1}$$

the standard logistic function. The bag is positive as long as one of the examples in the bag is positive; hence, the probability that the bag is positive is a "noisy OR" (Heckerman, 1989; Mason et al., 2000):

$$p_i = 1 - \prod_{j=1}^{J_i} (1 - p_{ij}) \tag{3.2}$$

Under this model the likelihood assigned to a set of training bags is as follows:

$$L^T = \prod_i p_i^{z_i} (1 - p_i)^{(1-z_i)}, \tag{3.3}$$

where $z_i \in \{0, 1\}$ is the label of bag i. The log likelihood is thus:

$$\log L^T = \sum_i \left[z_i \log p_i + (1 - z_i) \log(1 - p_i) \right]. \tag{3.4}$$

Following the AnyBoost approach, the weight on each example is given as the derivative of the cost function with respect to a change in the score of the example. The derivative of the log likelihood is:

$$
\begin{aligned}
\frac{\partial \log L^T}{\partial y_{ij}} &= \frac{\partial \log L^T}{\partial p_i} \cdot \frac{\partial p_i}{\partial p_{ij}} \cdot \frac{\partial p_{ij}}{\partial y_{ij}} \\
&= \frac{z_i - p_i}{p_i(1 - p_i)} \cdot \frac{1 - p_i}{1 - p_{ij}} \cdot (1 - p_{ij}) p_{ij} \\
&= \frac{z_i - p_i}{p_i} p_{ij}.
\end{aligned}
\tag{3.5}
$$

Each training example x_{ij} is given the weight

$$w_{ij}^T = \frac{\partial \log L^T}{\partial y_{ij}} = \frac{z_i - p_i}{p_i} p_{ij}. \tag{3.6}$$

Note, compared with the exponential weights in AdaBoost or logistic weights in LogitBoost that are always positive, the weights here are signed. For examples in the positive bags, $z_i = 1$, and the weight is positive. For negative examples, $z_i = 0$, and the weight is negative. For each round of AnyBoost, we search for a weak classifier that maximizes $\sum_{ij} f(x_{ij}) w_{ij}$, where $f(x_{ij})$ is the score assigned to the example by the weak classifier. For a binary classifier, $f(x_{ij}) \in \{-1, +1\}$. The selected weak classifier is the one that has the steepest decent for the log likelihood value $\log L^T$.

Examining Eq. (3.5), the weight on each example is the product of two quantities: the bag weight $w_{ij}^T(\text{bag}) = \frac{z_i - p_i}{p_i}$, and the instance weight $w_{ij}^T(\text{instance}) = p_{ij}$. Observe that $w_{ij}^T(\text{bag})$ for a negative bag is always -1. Therefore, the weight for a negative instance, p_{ij}, is the same that would result in a non-MIL AdaBoost framework (i.e. the negative examples are all equally negative). The weight on the positive instances is more complex. As learning proceeds and the probability of the bag approaches the target, the weight on the entire bag is reduced. Within the bag, the examples are assigned a weight that is higher for examples with higher scores. Intuitively, the algorithm selects a subset of examples to assign a higher positive weight, and these examples dominate subsequent learning.

The above feature selection step with AnyBoost can be performed very efficiently. The remaining task is to determine the parameter λ^t for the selected weak classifier. Unfortunately, there is no shortcut for this task. The target log likelihood function is very complex, and any changes in λ^t will impact all examples' p_{ij}, all bags' p_i, and, subsequently, $\log L^T$. In practice, usually a line search is adopted to maximize the target function. This becomes a challenging issue if the number of negative examples is huge, as for the case of learning a face detector. As described in Chapter 2, importance sampling is usually necessary in such cases. However, we observe that the line search process is sensitive to the percentage of examples being sampled. Our solution is to use as many examples as possible (limited by the amount of memory, computation time, etc.) to perform the line search, and shrink the resultant parameter λ^t in order to avoid overly aggressive step sizes. Training an object detector with the proposed MILBoost is thus usually much slower than with the boosting algorithm presented in Fig. 2.2.

A typical MILBoost training process is illustrated in Fig. 3.1.

3.1.2   ISR MILBOOST

The ISR criterion can also be used to derive a MILBoost algorithm. The criterion is less probabilistic, but the derivatives (and hence example weights) lead to a form of instance competition. Define $\psi_{ij} = \exp(y_{ij})$, $\Psi_i = \sum_{j=1}^{J_i} \psi_{ij}$ and

$$p_i = \frac{\Psi_i}{1 + \Psi_i}. \tag{3.7}$$

Keeler et al. (1990) argued that ψ_{ij} can be interpreted as the likelihood that the object occurs at x_{ij}. The quantity Ψ_i can be interpreted as a likelihood that some (at least one) instance is positive, and, finally, p_i is the probability that some bag is positive. Following the same derivation as Section 3.1.1, the example weights for the ISR framework are:

$$w_{ij}^T = \frac{\partial \log L^T}{\partial y_{ij}} = (z_i - p_i)\frac{\psi_{ij}}{\Psi_i} \tag{3.8}$$

Examining the ISR criteria reveals two key properties. The first is the form of the example weight, which is explicitly competitive. The examples in the bag compete for weight since the weight is normalized by the sum of the ψ_{ij}'s, or Ψ_i. Although the experimental evidence is weak, this rule perhaps leads to a very localized representation, where a single example is labeled positive and the other examples are labeled negative. The second property is that the negative examples also compete for weight. This turns out to be troublesome in the detection framework since there are many more negative examples than positive ones. In contrast, the Noisy OR criteria treats all negative examples as independent negative examples.

Input

- Training set $S = \{(x_{ij}, z_i), i = 1, \cdots, N, j = 1, \cdots, J_i\}$, where $z_i \in \{0, 1\}$. N_+ is the total number of positive bags. N_- is the total number of negative bags. $N_+ + N_- = N$.
- A ratio r between the number of positive and negative bags for future testing sets.
- T is the total number of weak classifiers, which can be set through cross-validation.

Initialize

- Take all positive bags and randomly sample Q negative bags.
- Initialize negative example weight factor $\eta_{\text{neg}} = \frac{N_+}{rQ}$. All examples are given an initial score $y_{\text{init}} = 0$.

MILBoost Learning
For $t = 1, \cdots, T$:

1. Compute weights for the sampled training set (Eq. (3.6)). For negative examples, multiply the computed weights with η_{neg}.
2. If necessary, use importance sampling to generate a small size training set, conduct feature pre-selection.
3. Select the weak classifier $f^t(\cdot)$ from the top features chosen during the previous feature pre-selection step, use line search to compute the corresponding step size λ^t.
4. Update scores of all examples in the medium size training set.
5. If t belong to set $A = \{2, 4, 8, 16, 32, \cdots\}$,
    - Update scores and weights of the whole training set using the previously selected weak classifiers.
    - Perform weight trimming to trim 10% of the negative bags.
    - Take all positive bags and randomly sample Q negative bags from the trimmed training set.
    - Update negative example weight ratio η_{neg} (Eq. (2.6)).

Output
Series of weak classifiers $f^t(\cdot)$ and the associated step sizes λ^t.

Figure 3.1: MILBoost learning with weight trimming and booststrapping.

Figure 3.2: Two example images with people in a wide variety of poses. The algorithm will attempt to detect *all* people in the images, including those that are looking away from the camera.

3.1.3   APPLICATION OF MILBOOST TO LOW RESOLUTION FACE DETECTION

Experimental results are performed using a set of 8 videos recorded in different conference rooms for a teleconferencing application. The goal is to detect low resolution faces in panoramic images. These images are acquired from a set of cameras near the center of the conference room, as shown in Fig. 3.2. A collection of 1856 images were sampled from the 8 videos. In all cases, the detector was trained on 7 video conferences and tested on the remaining video conference. There were a total of 12364 visible people in these images, each labeled by drawing a rectangle around the head of the person. The practical challenge is to steer a synthetic virtual camera toward the location of the speaker. The focus here is on person detection; determination of the person who is speaking will be discussed in Chapter 5.

In every training image, each person is labeled by hand. The labeler is instructed to draw a box around the head of the person. While this may seem like a reasonable geometric normalization, it ignores one critical issue, the *context*. At the available resolution (1056 × 144 pixels), the head is often less than 10 pixels wide. At this resolution, even for clear frontal faces, the best face detection algorithms frequently fail. There are simply too few pixels on the face. The only way to detect the head is to include the surrounding image context. Nevertheless, it is difficult to determine the correct quantity of image context, as the many possible normalized subwindows shown in Fig. 3.3.

Note even if the body context is used to assist in detection, it is difficult to foresee the effect of body pose. Some of the participants are facing right, others left, and still others are leaning far forward/backward (while taking notes or reclining). The same context image is not appropriate for all situations.

Both of these issues can be addressed with the use of MILBoost. Each positive head is represented, during training, by a large number of related image windows, as shown in Fig. 3.3. The MILBoost algorithm is then used to simultaneously learn a detector and determine the location and scale of the appropriate image context.

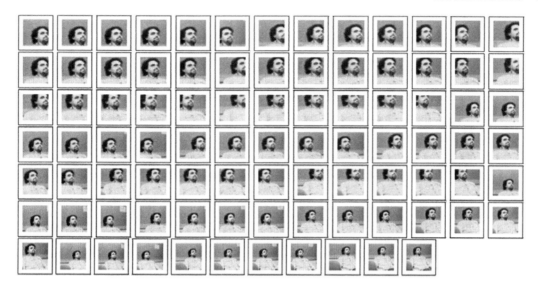

Figure 3.3: Some of the subwindows in one positive bag.

We follow the soft cascade learning approach summarized in Fig. 2.2 and Fig. 3.1 to train the detector. Learning is performed on a total of about 30 million subwindows in the 1856 images. Since the image resolution is very low, inspired by the work of Jones et al. (2003), we decided to add some additional feature images that captures the motion information.

The simplest motion filter is to compute the frame difference between two subsequent frames (Jones et al., 2003). When applied to our testing sequences, we find two major problems, demonstrated in Fig. 3.4. Fig. 3.4(a) and (b) are two subsequent frames captured in one of the recorded meetings. Person A was walking toward the whiteboard to give a presentation. Because of the low frame rate the detector is expected to run at, the difference image (c) has two big blobs for person A. Experiments show that the boosting algorithm often selects motion features among its top features, and such ghost blobs tend to cause false positives. Person B in the scene shows another problem. In a regular meeting, often someone in the room stays still for a few seconds; hence, the frame difference of person B is very small. This tends to cause false negatives.

To address the first problem, we use a simple three frame difference mechanism to derive the motion pattern. Let I_t be the input image at time t, we compute:

$$M_t = \min\left(|I_t - I_{t-1}|, |I_t - I_{t-2}|\right). \tag{3.9}$$

As shown in Fig. 3.4(d), Eq. (3.9) detects a motion region only when the current frame has large difference with the previous two frames, and can thus effectively remove the ghost blobs in Fig. 3.4(c). Note three frame difference was used in background modeling before (Collins et al., 2000).

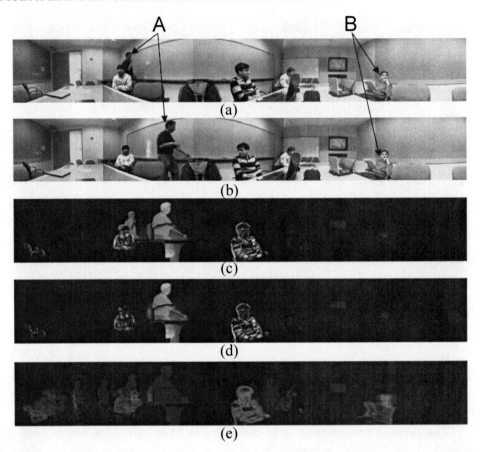

Figure 3.4: Illustration of motion filtering. (a) Video frame at $t-1$. (b) Video frame at t. (c) Difference image between (a) and (b). (d) Three frame difference image. (e) Running average of the three frame difference image.

We add another frame as Fig. 3.4(e), which is the running average of the three frame difference images:

$$R_t = \alpha M_t + (1-\alpha)R_{t-1}. \tag{3.10}$$

The running difference image accumulates the motion in the history, and it captures the long-term motion of people in the room. It can be seen that even though person B moved very slightly in this particular frame, the running difference image is still able to capture his body clearly.

Given the three frames I_t, M_t, and R_t, we use two kinds of simple visual features to train the classifier, as shown in Fig. 3.5. Note each detection window will cover the same location on all three images. The 1-rectangle feature on the left of Fig. 3.5 is computed on the difference image

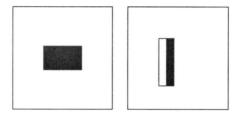

Figure 3.5: Example rectangle features shown relative to the enclosing detection window. Left: 1-rectangle feature; right: 2-rectangle feature.

and running difference image only. Single rectangle features allow the classifier to learn a data-dependent and location-dependent difference threshold. The 2-rectangle feature on the right is applied to all three images. This arrangement is to guarantee that all the features have zero-mean, so that they are less sensitive to lighting variations. For our particular application, we found adding more features such as 3-rectangle or 4-rectangle features gave very limited improvements on the classifier performance. A total set of 2654 rectangle filters are used for training. In each round, the optimal filter and threshold are selected. In each experiment, a total of 60 filters are learned.

We compared the classical AdaBoost with both noisy-OR MILboost and ISR MILBoost. For the MIL algorithms, there is one bag for each labeled head, containing those positive windows which overlap that head. Additionally, there is one negative bag for each image. After training, performance is evaluated on the held out conference video.

The experimental results comparing various learning algorithms are shown in Fig. 3.6. The horizontal axis is the false positive rate, defined as the number of false detections in each frame (all frames are of equal size 1056×144 pixels). The vertical axis is the false negative rate, defined as the portion of missed ground truth faces. During training, a set of positive windows are generated for each labeled example. All windows whose width is between 0.67 times and 1.5 times the head width and whose center is within 0.5 times the head width of the center of the head are labeled positive. An exception is made for AdaBoost, which has a tighter definition on positive examples (width between 0.83 and 1.2 times the head width and center within 0.2 times the head width) and produces better performance than the looser criterion. All windows that do not overlap with any head are considered negative. For each algorithm, one experiment uses the ground truth obtained by hand (which has small yet unavoidable errors). A second experiment corrupts this ground truth further, moving each head by a uniform random shift such that there is non-zero overlap with the true position. Note that conventional AdaBoost is much worse when trained using corrupted ground truth. Interestingly, Adaboost is worse than NorBoost using the "correct" ground truth, even with a tight definition of positive examples. We conjecture that this is due to unavoidable ambiguity in the training and testing data. Overall, the MIL detection results are practically useful. A typical example

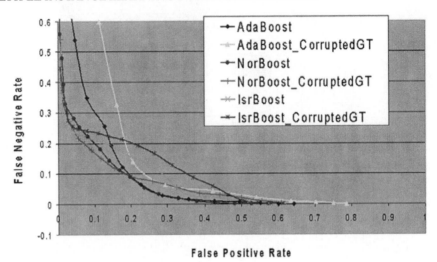

Figure 3.6: ROC comparison between various boosting rules.

Figure 3.7: One example from the testing dataset and overlaid results.

of detection results are shown in Fig. 3.7. Results shown are for the noisy OR algorithm. In order to simplify the display, significantly overlapping detection windows are averaged into a single window.

3.2 MULTIPLE CATEGORY BOOSTING

The MIL concept can be further extended to learning multiview object detectors. Since views of the same object can differ dramatically, the common practice in multi-view object detection has been "divide and conquer". Namely, the general class of objects is first divided into subcategories. Different classifiers can then be trained for different subcategories. For instance, faces can be categorized as frontal, left/right half profile, left/right profile, 0 degree in-plane rotation, ±30 degree in-plane rotation, etc. In the face detection work in (Li et al., 2002; Jones and Viola, 2003; Wu et al., 2004a), a pose estimator is first built to classify each example into one of the above subcategories. Each subcategory then trains its own classifier for detection with *manually* labeled data. The manual labeling process is certainly very labor-intensive, and it is sometimes difficult to do for tasks such as pedestrian detection or car detection. In (Seemann et al., 2006; Wu and Nevatia, 2007a), researchers proposed to obtain these labels via automatic clustering. Take the clustered boosted tree classifier in (Wu and Nevatia, 2007a) as an example. They applied a conventional k-means clustering algorithm to split the sample set into two parts when the learning rate slows down. They showed that by using the previously selected features for clustering, the learning algorithm converges faster and achieves better results.

One weakness that exhibits in early works (Li et al., 2002; Jones and Viola, 2003; Wu et al., 2004a) is the misclassification caused by the pose estimator. If a profile face is misclassified as frontal, it may never be detected in later stages. Huang et al. (2005) proposed vector boosting, which allows an example to be passed into multiple subcategory classifiers during testing, and the final results are fused through linear transform. Such a soft branching scheme can greatly reduce the risk of misclassification during testing.

Misclassification also happens in training. It could be caused by mislabeling. For instance, the boundary between frontal and half profile faces can be very subtle, and differs from person to person. For systems that rely on automatic clustering to derive the subcategory labels, misclassification can be very common. The misclassified examples appear as outliers in its designated subcategory, which may hinder the learning process and degrade the classifier performance. More importantly, although the manual labels or clustered results are meaningful for the system designer, there is no guarantee that they are optimal for learning the overall detector. Traditional training processes (Huang et al., 2005; Wu and Nevatia, 2007a) lack the flexibility to re-categorize examples during training, thus forming updated clusters that can help achieving the optimal performance.

In this section, we present *multiple category learning* (MCL), which overcomes the above issue through soft labeling. In essence, we maintain a set of likelihood values that describe the probability of each example belonging to each subcategory during training. These likelihood values are combined to compute the probability of the example being a positive example. The learning algorithm then maximizes the overall probability of all examples in the training data set. We demonstrate the MCL framework by two multiple category boosting (McBoost) algorithms: the *Probabilistic McBoost* (Prob-McBoost) and the *Winner-take-all McBoost* (WTA-McBoost). Prob-McBoost is derived based on the AnyBoost approach of Mason et al. (2000). It highly resembles the formulation of MilBoost above, which reveals the strong relationship between MCL and MIL. WTA-McBoost, on the other hand, is a more practical approach for multi-view object detection, where detection speed is a concern and attention cascade is a necessity. More specifically, we derive WTA-McBoost based on the assumption that the final classification of an example will only be determined by the highest probability in all subcategories, i.e., the winner will take all. The algorithm uses confidence-rated prediction (Schapire and Singer, 1999) with asymmetric cost and is thus very efficient to train and test.

3.2.1   PROBABILISTIC MCBOOST

Consider a two-class classification problem as follows. A set of labeled examples $\mathcal{S} = \{(x_i, z_i), i = 1, \cdots, N\}$ are given for training, where $z_i = 1$ for positive examples and $z_i = 0$ for negative examples.

In order to perform "divide and conquer", let us assume that the positive examples can be classified into $k = 1, \cdots, K$ subcategories, either by manual labeling or automatic clustering. Since the manual labels or the clustering results are not directly optimized for the overall two-class classification task, it would be suboptimal if we simply train K classifiers separately.

In our approach, we will train K boosting classifiers *jointly*. Recall in boosting each example is classified by a linear combination of weak classifiers. Let $y_{ik} = F_k^T(x_i) = \sum_t \lambda_k^t f_k^t(x_i)$ be the weighted sum of weak classifiers for subcategory k, often referred as the *score* of classifier k for example x_i. The probability of an example x_i belonging to subcategory k is given by:

$$p_{ik} = \frac{1}{1 + \exp(-y_{ik})}. \tag{3.11}$$

Using the Noisy OR generative model (Heckerman, 1989), we claim that the probability of an example x_i being a positive example is:

$$p_i = 1 - \prod_k (1 - p_{ik}). \tag{3.12}$$

The likelihood assigned to the whole training set S is thus:

$$L^T = \prod_i p_i^{z_i} (1 - p_i)^{1 - z_i}. \tag{3.13}$$

Following the AnyBoost approach (Mason et al., 2000), the weight on example x_i for subcategory k is given as the derivative of the cost function with respect to a change in the score of the example. The derivative of the log likelihood is:

$$\frac{\partial \log L^T}{\partial y_{ik}} = \frac{z_i - p_i}{p_i} p_{ik}. \tag{3.14}$$

For the classifier k, each training example x_i is given the weight

$$w_{ik}^T = \frac{z_i - p_i}{p_i} p_{ik}. \tag{3.15}$$

Note for negative examples, the weight $w_{ik} = -p_{ik}$, which is independent between different subcategories. Hence, during training we simply maintain K different weights for the negative examples and update them independently. In contrast, the weights w_{ik} of positive examples depend on the subcategory probability p_{ik} as well as the overall probability p_i. This is an intuitive result. If a positive example has high probability of belonging to a certain subcategory, its weights for the other subcategory classifiers shall drop since it will be classified correctly at the end.

The formulation of Prob-McBoost was independently proposed by Kim and Cipolla (2008) and Babenko et al. (2008). Readers may immediately notice the similarity between Prob-McBoost and noisy-OR-based MILBoost. Indeed, the weight calculation formula Eq. (3.14) is identical to that in MILBoost. Nevertheless, MIL and MCL address very different problems. In MIL, a bag of examples is classified as positive as long as one of the examples is positive. MIL is often used to identify common objects in bags that have uncertainty in location or scale. In MCL, an example is classified as positive as long as one of the subcategory classifiers classifies it as positive. MCL is more suitable for complex learning problems where it is desirable to cluster examples into different subcategories in order to improve the learning efficiency. Furthermore, in MIL a *single* classifier is learned throughout the process, while in MCL *multiple* classifiers are learned jointly.

During testing, examples are classified by comparing its overall probability, p_i as in Eq. (3.12), to a final threshold th. If $p_i > th$, this is a positive example; otherwise, it is a negative example. The computation of p_i requires the output of all the subcategory classifiers, p_{ik}, and is thus not suitable for early rejection (though if p_i is not calculated and each subcategory classifier makes decision individually, the intermediate thresholds can still be set through, say, the multiple instance pruning algorithm in Section 2.3). During training, similar to MilBoost, Prob-McBoost requires a line search to find the optimal weight λ_k^t in each subcategory classifier, which makes it slow to train and hard to deploy feature sharing (Torralba et al., 2007). In the next subsection, we introduce a much more practical scheme called Winner-take-all McBoost, which is faster to train, easier to share features among subcategory classifiers and more compatible with early rejection.

3.2.2   WINNER-TAKE-ALL MCBOOST

Let us again consider a set of training examples as $\mathcal{S} = \{(x_i, z_i), i = 1, \cdots, N\}$, where $z_i = 1$ for positive examples and $z_i = 0$ for negative examples. Assume the positive examples can be divided into K subcategories. For each subcategory, a boosting classifier will be learned. Let the score be $y_{ik}^T = F_k^T(x_i) = \sum_{t=1}^{T} \lambda_k^t f_k^t(x_i)$, where T is the number of weak classifiers in each subcategory classifier. In WTA-McBoost, we assume the highest score of all subcategories will be used to determine the fate of a given example. More specifically, let

$$y_i^T = \max_k y_{ik}^T. \tag{3.16}$$

Example x_i is classified as a positive example if y_i^T is greater than a threshold. Otherwise, it is a negative example. Following Friedman et al. (1998), we define the boosting loss as:

$$L^T = \sum_{i=1}^{N} [I(z_i = 1)\exp\{-y_i^T\} + I(z_i = 0)\exp\{y_i^T\}], \tag{3.17}$$

where $I(\cdot)$ is the indicator function. According to the statistical interpretation given by Friedman et al. (1998), minimizing this loss via boosting is equivalent to a stage-wise estimation procedure for fitting a cost-sensitive additive logistic regression model. In addition, as shown in (Schapire and Singer, 1999), the above loss function is an upper bound of the training error on the data set $\mathcal{S}$.

Unfortunately, minimizing the loss function in Eq. (3.17) is difficult and can be very expensive in computation. Notice

$$\exp\{y_i^T\} = \exp\{\max_k y_{ik}^T\} \leq \sum_k \exp\{y_{ik}^T\},\tag{3.18}$$

we instead optimize a looser bound as:

$$L^T = \sum_{i=1}^N \left[I(z_i = 1) \exp\{-y_i^T\} + I(z_i = 0) \sum_k \exp\{y_{ik}^T\} \right].\tag{3.19}$$

Since the subcategories of the positive examples are different from each other, it is unlikely that a negative example having a high score in one subcategory will have high score in another category. Hence the looser bound in Eq. (3.19) shall still be reasonably tight.

In the following, we devise a two-stage algorithm to minimize the asymmetric boost loss in Eq. (3.19). With weak classifiers at stage t, define the current *run-time label* of positive example x_i as:

$$l_i^t = \arg\max_k y_{ik}^t.\tag{3.20}$$

Based on these labels, we can split the loss function into K terms, $L^t = \sum_{k=1}^K L_k^t$, where

$$L_k^t = \sum_{i=1}^N \left[I(l_i^t = k) I(z_i = 1) \exp\{-y_{ik}^t\} + I(z_i = 0) \exp\{y_{ik}^t\} \right],\tag{3.21}$$

In the first stage of the algorithm, we assume the run-time labels are fixed, and search for the best weak classifiers $h_k^{t+1}(\cdot)$ and votes λ_k^{t+1} that minimize $\tilde{L}^{t+1} = \sum_{k=1}^K \tilde{L}_k^{t+1}$, where

$$\tilde{L}_k^{t+1} = \sum_{i=1}^N \left[I(l_i^t = k) I(z_i = 1) \exp\{-y_{ik}^{t+1}\} + I(z_i = 0) \exp\{y_{ik}^{t+1}\} \right].\tag{3.22}$$

This stage can be accomplished by performing boosting feature selection and vote computation for each subcategory *independently*. For instance, one can adopt the MBHBoost scheme proposed by Lin and Liu (2005), which trained multiple classes simultaneously and shared features among multiple classifiers. Since the boosting loss is convex, it is guaranteed that this boosting step will reduce the loss function, i.e., $\tilde{L}_k^{t+1} \leq L_k^t$, and $\tilde{L}^{t+1} \leq L^t$.

In the second stage, we update the run-time labels, namely:

$$l_i^{t+1} = \arg\max_k y_{ik}^{t+1}.\tag{3.23}$$

The loss function is updated as $L^{t+1} = \sum_{k=1}^K L_k^{t+1}$, where

$$L_k^{t+1} = \sum_{i=1}^N \left[I(l_i^{t+1} = k) I(z_i = 1) \exp\{-y_{ik}^{t+1}\} + I(z_i = 0) \exp\{y_{ik}^{t+1}\} \right].\tag{3.24}$$

Input

- Training examples $\{(x_i, z_i, s_i), i = 1, \cdots, N\}$, where $z_i \in \{0, 1\}$ for negative and positive examples. $s_i \in \{1, \cdots, K\}$ is the initial subcategory ID. For positive examples, s_i can be manually assigned or obtained through clustering. For negative examples, s_i can be randomly assigned (it will be ignored in WTA-McBoost).
- T is the total number of weak classifiers, which can be set through cross-validation.
- P is the label switching frequency.

Initialize

- Assign initial scores for all examples $y_{ik}^0 = 0$, where $k = 1, \cdots, K$ is the index of the subcategory classifier.
- Assign run-time label $l_i^0 = s_i$.

WTA-McBoost Learning

For $t = 1, \cdots, T$:

- According to the current run-time labels, perform confidence-rated boosting for each subcategory (Eq. (3.22)).
- If $\mod (t, P) = 0$, perform label switching for the positive examples ($z_i = 1$) according to Eq. (3.23).

Output

A set of K boosting classifiers $F_k^T(x_i) = \sum_{t=1}^{T} \lambda_k^t f_k^t(x_i)$.

Figure 3.8: WTA-McBoost learning.

It is straightforward to see that $L^{t+1} \leq \tilde{L}^{t+1}$; hence, both stages of the algorithm will reduce the loss function. Given that the asymmetric boost loss in Eq. (3.19) is non-negative, the algorithm is guaranteed to converge to a (local) minimum.

The run-time labels in WTA-McBoost can be updated after each weak classifier is added. In practice, it may be beneficial to update them less frequently to avoid label oscillation. Fig. 3.8 shows the detailed stages of WTA-McBoost. In this flowchart, the run-time-labels are updated every P weak classifiers are learned for each subcategory. A typical value of P is 32. Moreover, we may choose a large P at the very first round. This allows the subcategory classifiers to have a "burn in" period where they learn the general property of each subcategory. In our implementation, we start label switching after 96 weak classifiers are learned for each subcategory. Since label switching involves little computation, and it is done infrequently, the additional computational cost of WTA-McBoost, compared with the traditional approach of training each sub-class separately, is negligible.

Updating the run-time labels allows the positive example clusters to be re-formed during training, which can improve the final classification performance. In contrast, although clustering was used in (Wu and Nevatia, 2007a) during sample set splitting, their clusters are fixed and do not change during feature selection. This may hinder the learning process due to misclassification during clustering. On the other hand, note here we do not discuss subcategory splitting or merging. We may use manual labels to assign the initial subcategories, or we may use the splitting criteria and clustering method in (Wu and Nevatia, 2007a). In either case, WTA-McBoost can be applied to combat the misclassification issue and improve the overall classification performance.

3.2.3   EXPERIMENTAL RESULTS

We first test the WTA-McBoost algorithm on a multi-view face detection problem in its most straightforward fashion. A total of 100,000 face images with size 24×24 pixels are collected from various sources including the web, the Feret database (Phillips et al., 2000), the BioID database (Jesorsky et al., 2001), the PIE database (Sim et al., 2003), etc. These faces are manually labeled into 5 subcategories: frontal, left half profile, left profile, right half profile and right profile. Each subcategory contains 20,000 faces, with -10 degree to +10 degree in-plane rotation. The negative image set is also collected from the web, which contains about 1.2 billion image patches of size 24×24 pixels.

We ran a simple k-means clustering algorithm on the face images, where the number of clusters is given as 5. The distance between two facial images is measured as the Euclidian distance between their down-sampled images (12×12 pixels). The initial means of the clusters are computed by averaging the images with the same manual labels. The k-means clustering algorithm converges in about 50 iterations.

Our experiment compares the learning performance of WTA-McBoost, and that of the traditional approach, which trains each subcategory separately. In fact, if we skip the label switching step in WTA-McBoost (Eq. (3.23)), we have an implementation of the traditional approach. In both cases, a total of 800 weak classifiers are learned for each category, with *shared* Haar features and shared feature partitions. Note feature sharing may speed up the detection speed and improve generalizability, but it is not required by either approach. Fig. 3.9 shows the learning performance on the *training* data set. We compare the two approaches using the k-means clustered results as initial subcategory labels and using the manual labels as initial subcategory labels.

From Fig. 3.9, it can be seen that WTA-McBoost outperforms the traditional approach on the training data set. The horizontal axis is the false positive rate defined as the portion of negative examples that are classified as positive; the vertical axis is the detection rate defined as the percentage of positive examples correctly classified. Note when clustering is used as label initialization, the WTA-McBoost can reduce the false positive rate (FPR) by over 60%. The difference for manual labels is not as significant, with WTA-McBoost reducing the FPR by around 10%. Considering manual labels are generally much more meaningful (especially for faces) than clustered labels, the smaller reduction is expected. Fig. 3.10 shows the percentage of positive examples that switch labels

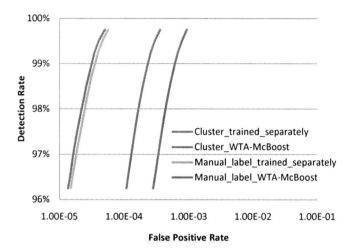

Figure 3.9: *Training* performance with WTA-McBoost and the traditional approach (each subcategory is trained separately), using the k-means clustered results as initial subcategory labels and using the manual labels as initial subcategory labels. Note the horizontal axis is in logarithmic scale.

during WTA-McBoost. Note as mentioned earlier, label switching starts at 96 weak classifiers, and it is done once every 32 weak classifiers afterwards. It can be seen that during the first few rounds, many examples switched labels. In particular, at 96 weak classifiers, as many as 34.47% of the positive examples have switched labels for clustering-based initialization, and 7.10% for manual-label-based initialization. The number drops very quickly. At 480 weak classifiers, only 2.43% positive examples switched labels for clustering-based initialization, and 0.72% for manual-label-based initialization. This quick convergence of run-time labels can be very useful. For instance, once there are very few positive examples that will switch labels, a test example at this stage can be safely classified into one of the subcategories, and a single subcategory classifier can be run afterwards, which saves a lot of computation (see Section 3.2.4 for more details).

It is interesting to examine the training examples that switch their labels after WTA-McBoost. Fig. 3.11 shows a few such examples when the subcategory labels are initialized manually. In the first row, the examples all have very extreme lighting conditions. Such examples are abundant since we included the PIE database (Sim et al., 2003) in our training. We found that many of these examples have switched their labels after WTA-McBoost. The new labels are consistent in that when the lights are from the left, the examples tend to be relabeled as left profile, and when the lights are from the right, the examples tend to be relabeled as right profile. It appears that for these examples with extreme lighting, categorizing them to one of the profile subcategories help improve their detection accuracy. In the second row, we show some examples where the new labels are different from the manual label but very close. Such examples are also plenty. These examples show the unclear

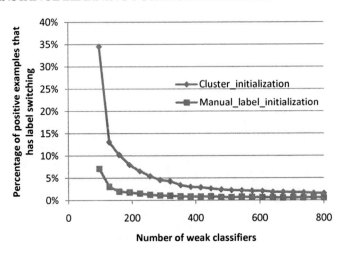

Figure 3.10: Percentage of positive examples that switch labels during WTA-McBoost.

boundary between neighboring subcategories, and it is often hard for human to be certain which subcategory the examples shall be assigned. The third row shows a few examples where the new labels after WTA-McBoost do not seem to make much sense. Lucky, there are less than 50 such examples in the total set of 100,000 face images.

Finally, we test the learned detectors on two standard data sets that are never seen during training, the CMU+MIT frontal data set (Rowley et al., 1998), and the CMU profile data set (Schneiderman and Kanade, 2000). It is worth mentioning that the latter data set contains many faces that have more than ±10 degree in-plane rotation, which are not represented in our training examples and *not* excluded in our experiments.

Fig. 3.12 shows the detector performance on the above two standard data sets. We also include a detector that is trained without subcategories with the learning framework in (Zhang and Viola, 2007), i.e., faces of all poses are mixed together and a single boosting classifier is trained for face/non-face classification (see also Chapter 2). The single boosting classifier uses the same Haar features and contains 2048 weak classifiers. A few observations can be made from Fig. 3.12. First, "divide and conquer" does help improve the performance. Even with the very naïve clustering-based initialization, and all subcategories are trained separately, "divide and conquer" still outperforms the single boosting classifier trained with all poses mixed. Second, WTA-McBoost can improve the detector performance significantly compared with the traditional approach, even with manual labels as initialization. For instance, on the CMU+MIT frontal data set, at 85% detection rate, WTA-McBoost reduces the number of false detections by 37% for clustering-based initialization; at 90% detection rate, WTA-McBoost reduces the number of false detections by 25% for manual-label-based initialization. Moreover, as mentioned earlier, WTA-McBoost requires negligible additional computation

Manual label	F	F	F	F	LHP	LHP	F	F	LHP	RHP
After WTA-McBoost	LP	LP	RP	RP	RP	RP	RHP	LHP	LP	LP
Manual label	F	F	LP	LP	RP	RP	LHP	LHP	RHP	RHP
After WTA-McBoost	LHP	LHP	LHP	LHP	RHP	RHP	LP	LP	RP	RP
Manual label	RP	LP	RP	RHP	RP	LHP	F	RHP	F	LHP
After WTA-McBoost	F	F	LHP	LHP	RHP	RHP	LP	LP	RP	RP

Figure 3.11: Training examples that switch their labels after WTA-McBoost. F: Frontal; LHP: Left Half Profile; RHP: Right Half Profile; LP: Left Profile; RP: Right Profile.

cost over the traditional approach; hence, we recommend that WTA-McBoost shall always be used for training "divide and conquer" style multi-view object detectors.

Another interesting observation from Fig. 3.12 is that detectors trained with manual-label-based initialization generally outperforms the naïve clustering-based initialization. The WTA-McBoost algorithm is a greedy adaptive labeling algorithm. Similar to other greedy searching algorithms such as k-means clustering, the performance of the trained detector can vary given different initial labels, and good initial labels are always helpful in getting a good classifier. In practice, the initial labels are often given manually or automatically through clustering, in which case WTA-McBoost will almost always guarantee to derive a better classifier than the traditional approach of training each subcategory separately.

3.2.4 A PRACTICAL MULTI-VIEW FACE DETECTOR

Although features can be shared among the multiple classifiers learned with WTA-McBoost, the computational cost for classifying a test window is still relatively high for real-world applications. For instance, the time spent on running a 5-category WTA-McBoost classifier with feature sharing is about 3 times more than with a single category classifier. To improve the running speed, we propose to adopt a three-layer architecture for multi-view face detection, as shown in Fig. 3.13. More specifically, a single category classifier is first trained, which includes faces at all different poses. Although according to Fig. 3.12, a single category classifier trained with all poses may perform sub-optimally, this layer is critical in improving the detection speed. The second layer of the classifier is

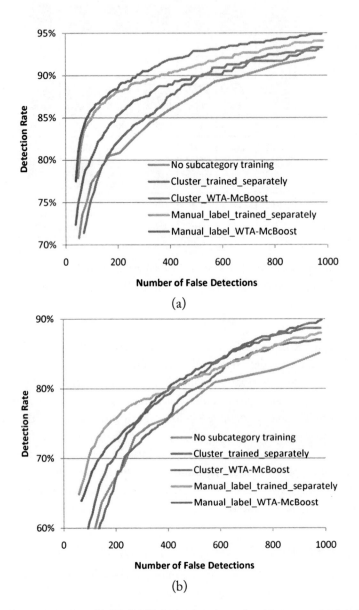

Figure 3.12: (a) Performance on CMU+MIT frontal data set (125 images, 483 labeled frontal faces). (b) Performance on CMU profile data set (208 images, 441 labeled faces with various poses, 73 (16.6%) of the faces have more than ±10 degree in plane rotations).

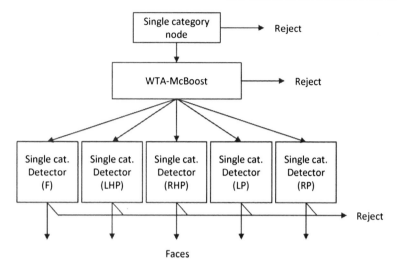

Figure 3.13: The three-layer architecture for multi-view face detection. Note in the third layer, the cluster IDs such as F, LHP, RHP, LP and RP are the output of WTA-McBoost and may not represent the true pose of the test window.

trained with WTA-McBoost, which allows the training positive examples to switch their subcategory labels during learning. As shown in Fig. 3.9, after a certain number of weak classifiers, the positive example clusters will converge, and the percentage of positive examples that switch labels during WTA-McBoost will be close to none. Once such a state has been achieved, we stop the WTA-McBoost learning and train separate single category classifiers for each cluster.

During testing, a test window is first passed into the first layer single category classifier. With less than 100 weak classifiers, the classifier is often capable of removing about 80-90% of the negative windows. If the test window is not rejected by the first layer, it will be passed into the WTA-McBoost classifier. The second layer may reject another 60-80% of the remaining negative training examples. At the end of WTA-McBoost, the test window will be given a cluster ID based on the highest score of the multiple subcategory classifiers, which is then used to determine which third layer classifier will be run. This branching is safe because WTA-McBoost has already converged at this stage.

We trained a 15-category multi-view face detector with the above architecture. The positive examples were again collected from the web, with artificially generated in-plane rotations from -45 to 45 degrees. These examples are categorized into 15 categories, similar to that in (Wu et al., 2004a). The negative example set was expanded to about 40 billion image patches. The first layer contains

128 weak classifiers, the WTA-McBoost-based classifier contains 160 weak classifiers[1], and the third layer classifiers contains 1152 weak classifiers each.

Fig. 3.14 compares the performance of our detector with a few existing approaches in the literature. It can been seen that on the CMU+MIT frontal data set, our detector's performance is comparable to many state-of-the-art algorithms. On the CMU profile data set, we perform worse than Wu et al. (2004a). There are a number of explanations. First, it could be attributed to the mismatch between the training data we have and the CMU profile data set. Second, we used the same original Haar feature sets as (Viola and Jones, 2001) for training. Extending this features set may lead to much better performance for profile face detection, as was reported in (Xiao et al., 2003; Huang et al., 2006). One possibility to improve the detector's performance is to design a more elaborate post filter to further remove false positive examples, such as those based on SVM or neural networks.

---

[1]We used 32 weak classifiers as the burn-in period for WTA-McBoost and switched labels every 8 weak classifiers. These settings are shorter than the examples we had in Section 3.2.3, but they still worked fine. The short burn-in period is mostly due to speed concerns.

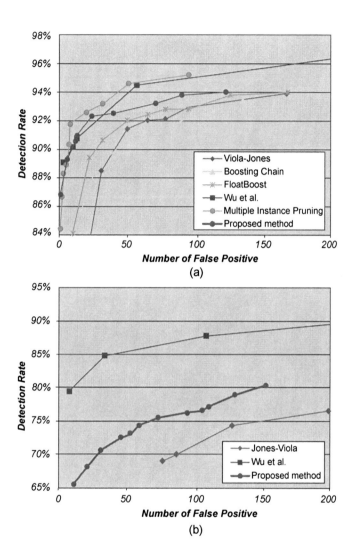

Figure 3.14: (a) Performance of our detector on CMU+MIT frontal data set (125 images, 483 labeled frontal faces). The algorithm is compared with Viola and Jones (2001), boosting chain (Xiao et al., 2003), FloatBoost (Li et al., 2002), Wu et al. (2004a) and multiple instance pruning (Zhang and Viola, 2007). (b) Performance on CMU profile data set (208 images, 441 labeled faces with various poses). The algorithm is compared with Jones and Viola (2003) and Wu et al. (2004a).

CHAPTER 4

Detector Adaptation

In the previous chapters, we have studied a number of algorithms that can be applied to train a classifier from a large set of labeled data. It is well-known that the performance of such a learned classifier will depend heavily on the representativeness of the labeled data used during training. If the training data contains only a small number of examples sampled in a particular test environment, the learned classifier may be too specific to be generalized to unseen data. On the other hand, if the training data is extensive, the classifier may generalize well but perform sub-optimally in a particular test environment.

In many situations, a generic classifier is needed to perform detection task for a variety of environments. For instance, when a person detector is applied in users' offices to detect people and set their presence status automatically, a generic detector is necessary so that the application can work in any office. On the other hand, although the test environment is unknown, the variation of the test environment is generally limited *after* the deployment of the classifier. That is, once the person detector has been deployed in the end user's office, the color of the walls, the furniture, the lighting condition, and even the people seen in the office remain largely the same. Ideally, we would like to have a mechanism for the generic classifier to adapt itself to the new environment and improve its performance over time.

In this chapter, we study the problem of detector adaptation, namely, how to adapt a generic classifier trained from extensive data-sets and improve its performance in a particular test environment. There are two main challenges in classifier adaptation. First, additional labeled data collected in the new test environment are often required for the adaptation. Since labeling is an expensive task for the end user, the amount of additional data has to be minimal. This is also the main obstacle to training a classifier from scratch only for that particular test environment. Second, a typical generic classifier may require thousands of example images or billions of sub-windows for training (e.g., the face detector in Chapter 2 and 3). It is usually impractical to send these data to the end user for a batch retraining of the classifier. One has to limit the amount of data associated with the generic classifier to be sent to the end user for classifier adaptation.

In various recognition tasks such as speech recognition and handwriting recognition, adaptation has become an indispensable tool to achieve high recognition accuracy. However, the importance of classifier adaptation for object detection tasks has rarely been explored in the literature. Recent developments of adaptation algorithms for tracking (Collins and Liu, 2003; Avidan, 2005; Li et al., 2007), in particular online-boosting-based methods (Grabner and Bischof, 2006; Liu and Yu, 2007), can be viewed as efforts toward classifier adaptation, but few have directly addressed the problem of adapting an *existing* generic detector to a particular test environment. In addition, most existing adaptation schemes are designed for a particular application, and applicable to a particular type of classifiers and a particular type of user labels. As a result, few can be easily applied to different applications with different classifiers or user labels.

We present a general formulation of the classifier adaptation problem and a novel Taylor-expansion-based adaptation method that is applicable to many classifiers. With our proposed method, the amount of data to be sent to the end user is only the gradient and Hessian of the classifier, which is typically very small. We demonstrate the application of the proposed method on two popular machine learning algorithms – logistic regression and boosting. Furthermore, in addition to the typical direct user labels that are given to individual examples, we introduce adaptation based on similarity labels, which are given to pairs of examples indicating whether they have the same, although unknown, labels. Similarity labels can be derived by an automatic tracking algorithm in video sequences, which creates a novel unsupervised classifier adaptation scheme that can be complementary to existing direct-label-based co-training schemes such as the work by Levin et al. (2003), Nair and Clark (2004) and Javed et al. (2005). Experiments are conducted on a challenging person detection task in conference rooms. We show that significant performance improvement can be achieved after adaptation from a generic person detector.

4.1   PROBLEM FORMULATION

4.1.1   PARAMETRIC LEARNING

Without loss of generality, consider a two-class classification problem as follows. A set of labeled examples $S = \{(x_i, z_i), i = 1, \cdots, N\}$ are given for training, where $z_i = 1$ for positive examples and $z_i = 0$ for negative examples. A parametric learning algorithm intends to find a mapping function

$$y = F(x|\Theta), \tag{4.1}$$

where y is the predicted label of the example, and Θ is a set of parameters for the mapping function. In order to find the optimal mapping parameters, a common practice is to define a cost function

$$C(F(x|\Theta), S) \tag{4.2}$$

over the training data-set. Various optimization algorithms can then be applied to minimize the cost function in order to obtain the optimal parameters Θ. For instance, gradient decent is one of the most popular algorithms used in machine learning. Numerous classifiers such as neural networks, logistic regression and boosting all have their own versions of gradient decent solution for cost function minimization (Bishop, 2006).

4.1.2   DETECTOR ADAPTATION

During adaptation, a generic classifier trained on an extensive data-set $S^{(o)}$ is given, denoted as $F(x|\Theta^{(o)})$, where superscript $^{(o)}$ indicates that the parameters are optimal on the *old* data. The user has collected data $S^{(n)}$ from a new test environment, where superscript $^{(n)}$ indicates *new*. The goal is to find a new set of parameters $\mathbf{\Theta}^{(n)}$, such that the classifier performs better in the new test environment.

While there are many possible ways to find $\mathbf{\Theta}^{(n)}$, here we propose to use:

$$
\begin{aligned}
\mathbf{\Theta}^{(n)} &= \underset{\mathbf{\Theta}}{\arg\min}\, J(\mathbf{\Theta}) \\
&= \underset{\mathbf{\Theta}}{\arg\min}\, C(F(x|\mathbf{\Theta}), \mathcal{S}^{(o)}) + \lambda D(F(x|\mathbf{\Theta}), \mathcal{S}^{(n)}),
\end{aligned}
\tag{4.3}
$$

where J is the revised overall cost function for adaptation; D is a cost function defined on the new data-set; and λ is a parameter controlling the relative importance of the old and the new data-set. Note D may be different from C because the labels on the new data-set may be in a different form (*e.g.*, Section 4.2.4).

Eq. (4.3) combines the cost functions on the old and the new data-sets, which ensures that the adapted classifier can work well even when the new data-set is very small. The remaining challenge is how to obtain $C(F(x|\mathbf{\Theta}), \mathcal{S}^{(o)})$. As mentioned earlier, the old data-set $\mathcal{S}^{(o)}$ is extensive and too huge to be made available for adaptation, which is performed at the end user's side. Our proposal is to use a compact representation or an approximation of the cost function on the old data-set to replace $C(F(x|\mathbf{\Theta}), \mathcal{S}^{(o)})$ in Eq. (4.3) during adaptation, *i.e.*:

$$
C(F(x|\mathbf{\Theta}), \mathcal{S}^{(o)}) \approx \tilde{C}(F(x|\mathbf{\Theta}), \mathbb{C}(\mathcal{S}^{(o)})),
\tag{4.4}
$$

where $\mathbb{C}(\mathcal{S}^{(o)})$ is a compact representation of the old data-set $\mathcal{S}^{(o)}$. Depending on the classifier's formulation, such a compact representation may or may not be easy to find. Below, we present a widely applicable approximation method based on Taylor expansion.

4.1.3   TAYLOR-EXPANSION-BASED ADAPTATION

We propose to use the Taylor expansion of the cost function on the old data as an approximation. For instance, with the second order Taylor expansion at the previously trained parameters $\mathbf{\Theta}^{(o)}$, we have:

$$
\begin{aligned}
C(F(x|\mathbf{\Theta})) &\approx C(F(x|\mathbf{\Theta}^{(o)})) + \\
&\quad \nabla C(F(x|\mathbf{\Theta}^{(o)}))(\mathbf{\Theta} - \mathbf{\Theta}^{(o)}) + \\
&\quad \frac{1}{2}(\mathbf{\Theta} - \mathbf{\Theta}^{(o)})^T \mathbf{H}_C(\mathbf{\Theta}^{(o)})(\mathbf{\Theta} - \mathbf{\Theta}^{(o)}),
\end{aligned}
\tag{4.5}
$$

where we represent parameter $\mathbf{\Theta}$ in a vector form and omit symbol $\mathcal{S}^{(o)}$ for conciseness. $\nabla C(F(x|\mathbf{\Theta}^{(o)}))$ is the gradient of the cost function, and $\mathbf{H}_C(\mathbf{\Theta}^{(o)})$ is the Hessian matrix whose elements comprise the second order derivative of the cost function with respect to $\mathbf{\Theta}$. With this approximation, the adaptation algorithm only needs to receive $\nabla C(F(x|\mathbf{\Theta}^{(o)}))$ and $\mathbf{H}_C(\mathbf{\Theta}^{(o)})$, which are generally much smaller in size compared with the original data-set $\mathcal{S}^{(o)}$.

Note the above Taylor expansion approximation is valid for smooth multivariate functions where $\nabla C(F(x|\Theta))$ and $\mathbf{H}_C(\Theta)$ exist within a *ball* in the space of Θ with center at $\Theta^{(o)}$ (Gill et al., 1981). The error of the approximation is on the order of $||\Theta - \Theta^{(o)}||^3$. In practice, we note many cost functions of parametric machine learning algorithms are indeed smooth around the optimized $\Theta^{(o)}$. In the following, we demonstrate the application of the above adaptation scheme on two popular machine learning algorithms: logistic regression and boosting.

4.2   ADAPTATION OF LOGISTIC REGRESSION CLASSIFIERS

4.2.1   LOGISTIC REGRESSION

Logistic regression is a very popular tool in machine learning (Bishop, 2006). In this method, a set of features $f_t(\cdot), t = 1, \cdots, T$ are extracted from the training data set $S = \{(x_i, z_i), i = 1, \cdots, N\}$. The likelihood of an example x_i being a positive example is:

$$p_i = \frac{1}{1 + \exp\{-\sum_{t=1}^{T} w_t f_t(x_i)\}},$$

(4.6)

where w_t is the set of parameters to be determined. The likelihood function of the whole data-set can be written as:

$$L^T = \prod_i p_i^{z_i} (1 - p_i)^{1-z_i}.$$

(4.7)

As usual, we can define a cost function by taking the negative logarithm of the likelihood, which gives the *cross-entropy* error function as:

$$C \triangleq -\frac{1}{N} \ln L^T = -\frac{1}{N} \sum_i \{z_i \ln p_i + (1 - z_i) \ln(1 - p_i)\}.$$

(4.8)

Logistic regression minimizes the above cost function on the training data-set to find the optimal set of parameters w_t. We refer the readers to (Minka, 2001) for a comparison between various algorithms to solve logistic regression.

4.2.2   ADAPTATION OF LOGISTIC REGRESSION CLASSIFIER

The gradient and Hessian of the logistic regression error function with respect to the parameters w_t can be easily computed as (Bishop, 2006):

$$\frac{\partial C}{\partial w_t} = \frac{1}{N} \sum_i (p_i - z_i) f_t(x_i),$$

(4.9)

$$\frac{\partial^2 C}{\partial w_t \partial w_s} = \frac{1}{N} \sum_i p_i (1 - p_i) f_t(x_i) f_s(x_i)$$

(4.10)

Denote $\mathbf{w} = [w_1, \cdots, w_T]'$ as the parameter vector; denote $\mathbf{p} = [p_1, \cdots, p_N]'$ and $\mathbf{z} = [z_1, \cdots, z_N]'$ as the likelihood and label vector of all examples; and denote $\mathbf{F}$ as the $N \times T$ design matrix with $f_t(x_i)$ as the $(i, t)^{\text{th}}$ element. Superscript $'$ stands for matrix transpose. We have in vector form:

$$\nabla C(\mathbf{w}) = \frac{1}{N} \mathbf{F}'(\mathbf{p} - \mathbf{z})$$
$$\mathbf{H}_C(\mathbf{w}) = \nabla \nabla C(\mathbf{w}) = \frac{1}{N} \mathbf{F}' \mathbf{R} \mathbf{F}, \tag{4.11}$$

where $\mathbf{R}$ is the $N \times N$ diagonal *weighting* matrix with elements $R_{ii} = p_i(1 - p_i)$.

It is therefore straightforward to apply Eq. (4.3) for logistic regression adaptation. Following the idea in Section 4.1.3, the cost function on the old data-set can be approximated as:

$$\begin{aligned} C(\mathbf{w}) \quad \approx \quad & C(\mathbf{w}^{(o)}) + \nabla C(\mathbf{w}^{(o)})(\mathbf{w} - \mathbf{w}^{(o)}) + \\ & \frac{1}{2}(\mathbf{w} - \mathbf{w}^{(o)})^T \mathbf{H}_C(\mathbf{w}^{(o)})(\mathbf{w} - \mathbf{w}^{(o)}), \end{aligned} \tag{4.12}$$

where $\nabla C(\mathbf{w}^{(o)})$ and $\mathbf{H}_C(\mathbf{w}^{(o)})$ are computed at the generic classifier's weight vector $\mathbf{w}^{(o)}$ on the old data-set. The cost function on the new data-set depends on the form of user labels. Next, we present two types of labels that can be used in our adaptation framework: direct labels and similarity labels. Note both types of labels can be obtained automatically, for example, via co-training (Nair and Clark, 2004; Roth et al., 2005) or tracking (Section 4.5.2).

4.2.3   DIRECT LABELS

Direct labels are labels given on examples directly, *e.g.*, $\mathcal{S}^{(n)} = \{(x_i^{(n)}, z_i^{(n)}), i = 1, \cdots, N^{(n)}\}$, where $x_i^{(n)}$ is the example and $z_i^{(n)}$ is the label information. Such labels are identical to those used for training the generic detector before adaptation. The cost function on the new data-set can be the same cross entropy error function defined in Eq. (4.8):

$$D \triangleq -\frac{1}{N^{(n)}} \sum_i \{z_i^{(n)} \ln p_i^{(n)} + (1 - z_i^{(n)}) \ln(1 - p_i^{(n)})\}, \tag{4.13}$$

where $p_i^{(n)}$ is defined as in Eq. (4.6). The overall cost function for classifier adaptation is hence:

$$\begin{aligned} J(\mathbf{w}) \quad = \quad & C(\mathbf{w}^{(o)}) + \nabla C(\mathbf{w}^{(o)})(\mathbf{w} - \mathbf{w}^{(o)}) + \\ & \frac{1}{2}(\mathbf{w} - \mathbf{w}^{(o)})^T \mathbf{H}_C(\mathbf{w}^{(o)})(\mathbf{w} - \mathbf{w}^{(o)}) + \\ & \lambda D(\mathbf{w}). \end{aligned} \tag{4.14}$$

We minimize the overall cost function by an efficient iterative technique based on the *Newton-Raphson* iterative optimization scheme. It takes the form:

$$\mathbf{w}^{[m+1]} = \mathbf{w}^{[m]} - \mathbf{H}_J^{-1}(\mathbf{w}^{[m]}) \nabla J(\mathbf{w}^{[m]}), \tag{4.15}$$

where m is the iteration index. It is easy to compute:

$$
\begin{aligned}
\nabla J(\mathbf{w}^{[m]}) &= \mathbf{H}_C(\mathbf{w}^{(o)})(\mathbf{w}^{[m]} - \mathbf{w}^{(o)}) + \\
&\quad \nabla C(\mathbf{w}^{(o)}) + \lambda \nabla D(\mathbf{w}^{[m]}), \\
\mathbf{H}_J(\mathbf{w}^{[m]}) &= \mathbf{H}_C(\mathbf{w}^{(o)}) + \lambda \mathbf{H}_D(\mathbf{w}^{[m]}),
\end{aligned}
\tag{4.16}
$$

where the gradient and Hessian of the error function $D(\mathbf{w})$ *on the new data-set* can be computed in the same way as in Eq. (4.11).

During the iterative optimization process, the weight vector of the generic classifier is used for initialization:

$$
\mathbf{w}^{[0]} = \mathbf{w}^{(o)}.
\tag{4.17}
$$

We iterate on Eq. (4.15) until the Newton decrement is less than a certain threshold ξ:

$$
\sqrt{\nabla J(\mathbf{w}^{[m]})' \mathbf{H}_J^{-1}(\mathbf{w}^{[m]}) \nabla J(\mathbf{w}^{[m]})} < \xi.
\tag{4.18}
$$

4.2.4   SIMILARITY LABELS

Instead of labeling examples with positive or negative tags, one may also specify similarity labels, which indicates whether two examples share the same direct label or not. Similarity label takes the form of $\mathcal{S}^{(n)} = \{(x_{i1}^{(n)}, x_{i2}^{(n)}, z_i^{(n)}), i = 1, \cdots, N^{(n)}\}$, where $x_{i1}^{(n)}$ and $x_{i2}^{(n)}$ are the two examples, $z_i^{(n)} = 1$ indicates the two examples should have the same label, and $z_i^{(n)} = 0$ indicates the two examples should have different labels. In the following, we skip the superscript $^{(n)}$ to simplify the notations.

The probability of x_{i1} and x_{i2} sharing the same label can be written as:

$$
p_i = p_{i1} p_{i2} + (1 - p_{i1})(1 - p_{i2}),
\tag{4.19}
$$

where

$$
p_{il} = \frac{1}{1 + \exp\{-\sum_t w_t f_t(x_{il})\}}, l \in \{1, 2\}.
\tag{4.20}
$$

The cross-entropy error function is again:

$$
D \triangleq -\frac{1}{N} \sum_i \{z_i \ln p_i + (1 - z_i) \ln(1 - p_i)\}.
\tag{4.21}
$$

We still resort to the Newton-Raphson method to find the optimal parameter vector as in Eq. (4.15) and (4.16), except that the gradient and Hessian of the cost function on the new data-set needs to be revised. The gradient of the cost function on the new data-set is:

$$
\frac{\partial D}{\partial w_t} = \frac{1}{N} \sum_i \frac{(p_i - z_i) g_{ti}}{r_i},
\tag{4.22}
$$

where

$$
\begin{aligned}
r_i &= p_i(1 - p_i), \\
g_{ti} &= \frac{\partial p_i}{\partial w_t} = -q_{i2} r_{i1} f_t(x_{i1}) - q_{i1} r_{i2} f_t(x_{i2}),
\end{aligned}
$$

with $r_{il} = p_{il}(1 - p_{il}), q_{il} = \frac{\partial r_{il}}{\partial p_{il}} = 1 - 2p_{il}, l \in \{1, 2\}$. Second order derivatives are:

$$
\frac{\partial^2 D}{\partial w_t \partial w_s} = \frac{1}{N} \sum_i \left\{ \frac{[p_i^2 + z_i q_i] g_{ti} g_{si}}{r_i^2} + \frac{(p_i - z_i) h_{tsi}}{r_i} \right\}, \tag{4.23}
$$

where

$$
\begin{aligned}
q_i &= \frac{\partial r_i}{\partial p_i} = 1 - 2p_i, \\
h_{tsi} &= \frac{\partial g_{si}}{\partial w_t} \\
&= 2r_{i1} r_{i2}[f_t(x_{i1}) f_s(x_{i2}) + f_t(x_{i2}) f_s(x_{i1})] - \\
&\quad q_{i1} q_{i2}[r_{i1} f_t(x_{i1}) f_s(x_{i1}) + r_{i2} f_t(x_{i2}) f_s(x_{i2})]
\end{aligned}
$$

Note the Hessian matrix for similarity labels on the new data-set is not necessarily positive definite; hence, optimizing the error function $D(\mathbf{w})$ on the new data-set alone does not ensure a global minimum. Fortunately, $\mathbf{w}^{(o)}$ is the global optimal estimate minimizing the error function on the old data-set and serves well as a good initial estimate in the optimization for adaptation.

4.3 ADAPTATION OF BOOSTING CLASSIFIERS

In a typical boosting classifier, each example is classified by a linear combination of weak classifiers. Given a test example x_i, define the *score* of the example s_k as a weighted sum of weak classifiers $h_j(\cdot)$, *i.e.*,

$$
s_i = \sum_t \lambda_t f_t(x_i) \tag{4.24}
$$

where $f_t(x_i)$ can be written as:

$$
f_t(x_i) = \begin{cases} +1 & \text{if } h_t(x_i) > H_t \\ -1 & \text{otherwise} \end{cases} \tag{4.25}
$$

where H_t is the threshold for feature $h_t(\cdot)$ or weak classifier $f_t(\cdot)$. The final decision is made by comparing the example's score with an overall threshold th. That is, if $s_i >$ th, then example x_i is a positive example; otherwise, x_i is a negative example.

As described in previous chapters, Friedman et al. (1998) showed that the AdaBoost algorithms are indeed Newton methods for optimizing a particular exponential loss function – a criterion which behaves much like the log-likelihood on the logistic scale. In (Mason et al., 2000), Mason *et al.* showed that boosting can be viewed as a gradient-decent algorithm in the function space. The probability of an example being positive can be written as:

$$p_i = \frac{1}{1 + \exp\{-s_i\}},$$
(4.26)

Subsequently, we can use gradient-decent to search for the weak classifiers $f_t(\cdot)$ and the weights λ_t with the same cost function as in Eq. (4.8) (Mason et al., 2000).

Under the above AnyBoost formulation, if we restrict ourselves to updating only the weights λ_t, the adaptation of a boosting classifier can be identical to the adaptation of the logistic regression classifier. The difference is that in logistic regression, the features $h_t(\cdot)$ are usually real-valued. In contrast, in a boosting classifier, the weak classifiers $f_t(\cdot)$ are binary. However, this is no impact on the application of adaptation on boosting classifiers.

In some recent work for tracking (Grabner and Bischof, 2006; Liu and Yu, 2007), the weak classifiers were also updated during tracking. One approach is to compute the gradient of the cost function over weak classifiers numerically, as was done in (Liu and Yu, 2007). Alternatively, one can record the compact information (e.g., the gradient and the Hessian) for all possible combinations of weak classifiers. This may result in a large amount of information to be sent to the end user; hence, its application for classifier adaptation may be limited.

4.4   DISCUSSIONS AND RELATED WORK

Classifier adaptation, as presented in the form of Section 4.1.2, can be viewed as a very general formulation of classifier learning. For instance, if we set the parameter λ in Eq. (4.3) very large, minimizing Eq. (4.3) is equivalent to training a classifier on the new data-set only. This may indeed be the best practice if sufficient amount of data are collected in the test environment. On the other hand, if the new data-set contains examples given one-by-one, and the adaptation algorithm runs for every new input example, the methodology shown in Section 4.1.2 can be used to explain many online learning or sequential learning algorithms.

Many recent approaches that involve online learning for detection/tracking (Collins and Liu, 2003; Avidan, 2005; Grabner and Bischof, 2006; Li et al., 2007; Liu and Yu, 2007; Pham and Cham, 2007a) are related to our work. However, we found some of these methods (e.g., (Avidan, 2005; Liu and Yu, 2007)) do not consider the cost function on the old data-set. Instead, they take the feature set from the previous frame, update it if necessary, and then learn a new weight vector solely based on the tracked result in the current frame. This approach can cause model drifting in tracking. Collins and Liu (2003) addressed this issue by using the first frame as an "anchor" frame, effectively always keeping a small old data-set in its original form. In (Li et al., 2007, 2008), the generic person detector is in fact never updated. The authors relied on some simple classifiers (e.g., LDA) to learn the appearance change of the tracked object and combine it with the generic person detector to avoid drifting. While this approach is interesting, we consider it a very special case of classifier adaptation that operates on the *fused* classifier instead of the generic person detector.

Another family of related work is co-training-based detector adaptation or improvement (Levin et al., 2003; Nair and Clark, 2004; Roth et al., 2005; Javed et al., 2005; Conaire et al., 2007), or more generally, semi-supervised learning (Zhu, 2007). In co-training, multiple independent classifiers are applied on the same examples. If some of the classifiers have high confidence on a particular example, it can be used to update the remaining classifiers. Therefore, co-training is a mechanism to obtain additional training examples from unlabeled data, and our proposed adaptation algorithm can be applied once these examples are obtained. As for the adaptation algorithms used after new examples are obtained, Nair and Clark (2004) used an online Winnow algorithm to update the classifier; Javed et al. (2005) and Roth et al. (2005) used online-boosting-based on the work in (Oza, 2002). Our proposed Taylor-expansion-based classifier adaptation algorithm can be applied to both, and many other classifiers such as linear discriminant analysis, neural networks, logistic regression, etc. In addition, our algorithm can handle data with similarity labels, which cannot be used for traditional online Winnow or boosting.

One paper addressing the same problem as ours is the work by Huang et al. (2007b). Instead of the second order Taylor expansion adopted here, they approximated the joint likelihood by the product of corresponding weighted marginal distributions on each weak classifier. One clear advantage of the Taylor-expansion-based algorithm is its capability to handle similarity labels. Nevertheless, strictly speaking, both approaches are approximations and have limitations. The approximation of the joint likelihood with its marginal distributions is problematic as in boosting learning the late features are dependent upon early features. Our Taylor expansion solution could be perfect if arbitrary orders of the expansion can be preserved. However, in practice, that will render the representation humongous and inappropriate for adaptation.

Interestingly, the Taylor-expansion-based adaptation can also be viewed as a regularization method for parametric learning. Take logistic regression as an example. Using the Bayesian formulation, it can be shown that a Gaussian or Laplacian prior of the parameter vector can lead to a logistic regression formulation with $\ell 2$ or $\ell 1$ regularization (Bishop, 2006). Our proposed method can be viewed as using the Hessian matrix on the old data-set to regularize the logistic regression optimization on the new data-set. It certainly belongs to the category of $\ell 2$ regularization. However, unlike the widely used i.i.d. Gaussian prior (Minka, 2001), the Hessian matrix from the old data-set contains more information about each parameter and their correlations; hence, it can serve better when the new data-set is small.

4.5   EXPERIMENTAL RESULTS

We test the proposed classifier adaptation algorithm on the task of detecting people from panoramic videos of conference rooms, as shown in Figure 4.1. The task is very challenging due to pose variations, occlusions, small head sizes (*e.g.*, as small as 10×10 pixels), non-static background (*e.g.*, moving chairs and monitor contents), lighting variation, *etc.* In Section 3.1, we have studied the same problem with the MILBoost algorithm for learning. In this section, we focus on how to perform detector adaptation.

Figure 4.1: Example panoramic views of two meeting rooms for person detection. In the upper image, the blue box is one of the hand-labeled ground truth heads. The red box is the expanded box that includes shoulder for detector training.

A generic boosting-based person detector is trained on 93 meetings (6350 labeled frames) collected in more than 10 different rooms. In every labeled frame, each person is marked by a hand-drawn box around the head of the person. The blue box in Figure 4.1 shows such an example. Since the head size can be very small, we expand the ground truth box to include people's shoulders, making it effectively an upper body detector (see the red box in Figure 4.1). The minimum detection window size is 35×35, and the 93 sequences comprise over 100 million examples (both positive and negative) for training.

In addition to the monochrome images, two additional feature images are used. One measures the difference between subsequent video frames and the other measures the long term average of the temporal difference frames (See Section 3.1.3 for more details). A set of 6946 Haar like features are extracted from these three images and serve as the feature pool for boosting. We adopt the logistic variation of Adaboost developed by Collins et al. (2002) to train the classifier. The resultant detector has a total of 120 weak classifiers.

The generic person detector performs well in unseen meeting rooms. However, since a meeting room often has limited variations in background and lighting, we believe a classifier adaptation algorithm can help improve the generic detector's performance further. We collected a total of 17 meetings in the same meeting room (upper image of Figure 4.1) during a period of one week. A total of 265 frames are labeled as before. The 17 meetings are randomly split into two groups: 8 meetings (128 labeled frames) for adaptation and 9 meetings (137 labeled frames which contain 527 persons in total) for testing. The adaptation is performed on the weights (λ_t in Eq. (4.24)) of the weak classifiers only, and the optimization is done through the Newton-Raphson scheme (Eq. (4.15) and (4.16)).

4.5.1   RESULTS ON DIRECT LABELS

In the first set of experiments, we assume the end user has provided direct labels for some of the frames in the adaptation data-set. Figure 4.2 shows the adapted detectors' receiver operating

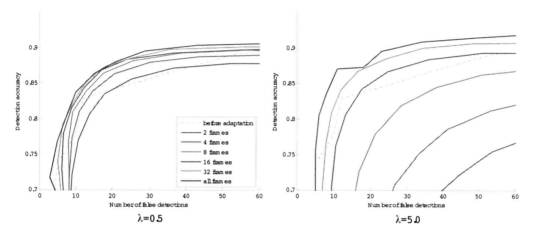

Figure 4.2: Adapted person detector performance with direct labels. In each figure, λ is fixed. The two figures share the same legend.

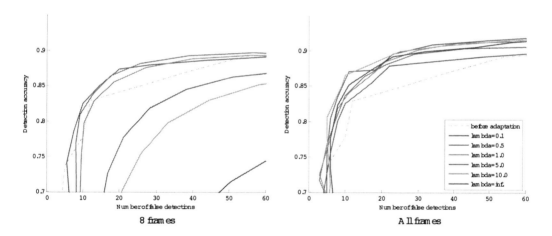

Figure 4.3: Adapted person detector performance with direct labels. In each figure, the number of labeled frames is fixed. The two figures share the same legend.

characteristic (ROC) curves on the 9 testing sequences for λ = 0.5 and 5.0. The horizontal axis is the number of false detections in the whole test set; the vertical axis is the detection accuracy defined as the percentage of true persons correctly detected. In each figure, we gradually increase the number of labeled frames in order to observe the impact of the amount of labeled data on the adaptation performance. Note each labeled frame contributes around 16,000 labeled patches for training. When all 128 frames are used for adaptation, the total number of labeled examples is over 2 million. For the curves that use N (where $N = 2, 4, 8, 16, 32$) labeled frames, we randomly sample N labeled frames from the adaptation data-set to perform classifier adaptation. Each ROC curve in Figure 4.2 is the *average* of 100 trials of such random sampling.

From Figure 4.2, it can be seen that in general the more labeled frames one has the better the performance is after adaptation. When $\lambda = 0.5$, with 4 newly labeled frames in the test environment, the adapted classifier already outperforms the generic detector. When all frames in the adaptation data-set are used, we observe the detection accuracy improves from 85.6% to 90.4% with 30 false detections ($\lambda = 5.0$), a decrease of around 33.3% in detection error.

Table 4.1: Optimal λ range for different amount of ground truth data. N is the number of labeled frames used for adaptation.

N	2	4	8	16	32	all
λ	≤ 0.1	≤ 0.1	0.1-0.5	0.5-1.0	0.5-1.0	5.0-10.0

Figure 4.3 shows two sets of curves when the number of labeled frames are 8 and 128 (all available frames). We vary the parameter λ from 0.1 to infinity. As mentioned earlier, when λ is infinite, classifier adaptation is equivalent to re-training a detector on the newly labeled data only. It can be seen that in both sets of curves, setting λ to infinity does not achieve the best performance. This demonstrates the necessity of having the Hessian of the generic classifier as the regularization term. In Table 4.1, we show the optimal λ range with respect to the number of labeled frames for adaptation. It can be seen that when the number of labeled frames is small, a small λ tends to give better results. On the other hand, if a large number of labeled frames are available, a large λ tends to perform better.

4.5.2   RESULTS ON SIMILARITY LABELS

We next report our experimental results on similarity labels. All 128 frames from the 8 adaptation sequences are used in the following experiments. These frames are organized into 120 pairs, where each pair consists of two subsequent frames in the *same* meeting sequence. Note, however, that a pair of subsequent frames may be a few seconds apart.

In the first experiment, we give ground truth similarity labels to the adaptation algorithm. For each pair of frames, we choose a detection window from one frame, and randomly choose another detection window from the other frame. The similarity label is computed from the actual direct labels of the two windows, i.e., 1 if they have the same direct label and 0, otherwise. This sampling process continues until all detection windows in the first frame have been selected. In total, around 2 million example pairs are created for training (most of them are pairs of negative examples).

The data collected above are then fed into the classifier adaptation algorithm as described in Section 4.2.4. Figure 4.4 shows the detector performance after adaptation. Note with all λ values, the adapted detector outperforms the generic detector (Figure 4.4(a)). Figure 4.4(b) gathers the best

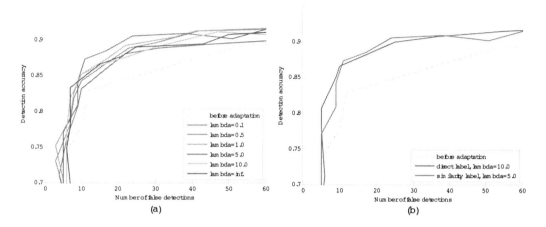

Figure 4.4: Adapted person detector performance with ground truth similarity labels. (a) Using all frames for adaptation, the performance of the adapted classifier with respect to different λ. (b) Best adaptation performance comparison using direct labels and similarity labels.

adapted detector with direct labels ($\lambda = 10.0$) and similarity labels ($\lambda = 5.0$). The two ROC curves are very similar; hence, similarity labeling is *as effective as* direct labeling when used for adaptation.

Finally, we present a simple algorithm to generate similarity labels automatically for adaptation. First, define the *histogram distance* between two detection windows as follows. Let the histogram of the first window be $\mathbf{p} = \{p(u), u = 1, \cdots, m\}$, where m is the total number of bins, and the histogram of the second window be $\mathbf{q} = \{q(u), u = 1, \cdots, m\}$. The distance between the two discrete distributions is defined as (Comaniciu et al., 2003):

$$d = \sqrt{1 - \rho[\mathbf{p}, \mathbf{q}]}, \tag{4.27}$$

where

$$\rho[\mathbf{p}, \mathbf{q}] = \sum_{u=1}^{m} \sqrt{p(u)q(u)} \tag{4.28}$$

is the sample estimate of the Bhattacharyya coefficient between $\mathbf{p}$ and $\mathbf{q}$.

For each pair of frames, we choose a detection window from one frame and search for a window (with identical size) in the other frame that has the smallest histogram distance to the first window. Once the window with the smallest distance has been found, we put the example pair in the training set with label 1 if the distance $d < 0.03$. Over 1 million example pairs are constructed automatically, among which 10 pairs have the wrong labels ($\sim 0.001\%$).

Figure 4.5 shows the adapted detector's performance when varying λ. Note the adapted detector outperforms the generic detector when $\lambda < 0.1$. The small λ values indicate that we shall rely more on the generic detector when the similarity labels are provided automatically by a rudimentary

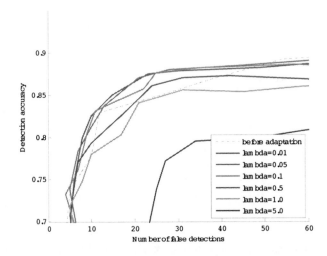

Figure 4.5: Adapted person detector performance with automatically generated similarity labels.

algorithm such as this one. We expect a more elaborated algorithm for generating the similarity labels may lead to further improvements in the detector performance.

A few challenges remain as our future work. For instance, while the rule of thumb is to use a small λ value when the amount of data collected in the new test environment is small (as shown in Table 4.1 based on our empirical study), determining the optimal λ is still a difficult task. Also, the adapted classifier may require a different final threshold. Typically, the threshold is derived from a validation data-set, but such a data-set may be difficult to obtain without significant user effort.

CHAPTER 5

Other Applications

We have focused on face detection almost exclusively in the previous chapters. In this chapter, we will present two other applications of boosting learning. These two applications extend the above algorithms in two ways: the learning algorithm itself, and the features being used for learning.

The first application is on face verification for filtering and ranking image/video search results on celebrities. Namely, given a small set of face images of a celebrity (e.g., the top query results from a text-based search engine), verify whether the celebrity is in other images and videos that are returned by the search engine. Although the problem sounds very different from face detection, the solution is actually similar. We propose an algorithm called boosted multi-task learning (MTL), which extends MILBoost with a graphical model. The main motivation is that the number of training images for each celebrity may be limited, and thus a typical machine learning algorithm may easily overfit on the training data set. Boosted MTL learns K boosting classifiers jointly for M celebrities, where $M \geq K$. Every celebrity then composes his/her own combined classifier, which is a mixture of the K classifiers. The key observation is that celebrities who have commonality in their features should be explained by the same boosting classifiers, while those who do not have commonality should have a different mixture of the K boosting classifiers. Boosted MTL effectively overcomes overfitting and achieves better results than training individual classifiers for each person or training a single classifier that simply tells if two images are from the same person or not, as shown in our experimental results.

The second application is on active speaker detection. The goal is to find who is speaking in a conference room, given a microphone array and a panoramic video of the room. Traditionally, sound source localization (SSL) and multi-person detection are conducted separately and fused with decision level fusion. Here we show that by combining audio and visual features in a boosting framework, we can determine the speaker's position more efficiently and more accurately. In experiments that includes hundreds of real-world meetings, the proposed method reduces the error rate of SSL-only approach by 24.6%, and the SSL and multi-person detection (MPD) decision level fusion approach by 20.9%.

5.1 FACE VERIFICATION WITH BOOSTED MULTI-TASK LEARNING

5.1.1 INTRODUCTION

There is an increasing interest in applying face recognition technologies to Internet search, especially verification of celebrities on the web. Current search engines mainly use text information to search for images and videos of celebrities. Although the top few returned examples are often satisfactory,

(a)

(b)

(c)

Figure 5.1: Examples of faces of celebrities from our data sets for experiments. (a) Face images of Bill Gates. (b) Faces of Bill Gates from videos. (c) Faces images of three other celebrities. Images in the same row belong to the same celebrity.

the precision drops quickly when more lower ranked examples are included. Apparently, only using text information is not enough to search for images and videos with high accuracy. For example, if "Eva Longoria" is input as key words to query from YouTube, the returned video ranked as No. 9 is actually showing how to fake a bob with the hair style of Eva Longoria, and Eva does not show up in that video at all. Furthermore, when users search for videos of a celebrity, they may also want to know when the celebrity appears in a video and the frequency of appearance, which is an indicator of how much a video is relevant to the celebrity. For example, people may be more interested in a MTV of Eva than a video of a long news program where Eva just shows up once. These tasks cannot be done with text information alone. So face recognition technologies are expected to help.

Given a small set of face images of a celebrity, our goal is to verify the celebrity from more images and videos on the web. With minor human intervention, a small training set of image examples can be obtained from the top of the results of text query. Faces recognition has achieved significant progress under controlled conditions. However, web-based face verification is a much harder problem since the web is an open environment (Huang et al., 2008; Verschae et al., 2008) where pose, lighting, expression, age and makeup variations are more complicated. Using images as training data to verify faces in videos has extra difficulties, since faces in videos are often of smaller resolutions, with more blurring effect, with larger pose variations, and under worse lighting conditions than faces in images. Some examples of faces in images and videos on the web are shown in Figure 5.17.

Web-based face recognition is an emerging interesting research topic. Yagnik and Islam (2007) used the text-image co-occurrence as a weak signal to learn a set of consistent face models from a very large and noisy training set of face images of celebrities on the web. Face images of a celebrity are clustered and outliers are removed. This consistency learning framework requires a large computational cost. Stone et al. (2008) utilized the social network context provided by Facebook to autotag personal photographs on the web.

Boosting was applied to face recognition using PCA, LDA and Gabor features to build weak classifiers (Guo et al., 2001; Lu et al., 2006; Yang et al., 2004). The work most relevant to ours is (Zhang et al., 2004), where histograms of local binary patterns (LBP) inside local regions are used to build the weak classifiers in AdaBoosting. Although it was better than other boosted features, the improvement compared with directly using LBP in (Ahonen et al., 2004) is marginal. In our work, counts of bins instead of local histograms are used to build weak classifiers. Compared with (Zhang et al., 2004), we have a much larger pool of more independent features and weak classifiers are simpler. Our experimental results show that this leads to a surprising big improvement on face verification accuracy and makes verification hundreds of times faster compared with both (Ahonen et al., 2004) and (Zhang et al., 2004). All these boosting algorithms train classifiers for different individuals separately or a generic classifier applying to all the faces. When each individual only has a small training examples, they have to face the problems explained earlier.

5.1.2   ADABOOSTING LBP

LBP is a powerful texture descriptor introduced by Ojala et al. (2002). As shown in Figure 5.2 (a), it defines the neighborhood of a pixel i by uniformly sampling P pixels along the circle centered at i with radius R. If the pixels in the neighborhood do not fall exactly on the grid, their values are estimated using bilinear interpolation. The pixels in the neighborhood are assigned with binary numbers (0 or 1), depending on whether the value of that pixel is smaller than the value of pixel i as shown in Figure 5.2 (b). Pixel i is labeled as one of the local binary patterns by converting the binary array into a decimal number. For instance, in Figure 5.2 (b), the centered pixel will be labeled as $11100000 = 224$.

One extension of LBP is to only keep *uniform patterns* while mapping all the *non-uniform patterns* to a single label. A local binary pattern is deemed uniform if it contains at most two bitwise transitions from 0 to 1 or vice versa when the binary string is considered circular. For example, 00000000, 00011110 and 10000011 are uniform patterns, and 00010100 is not. As observed

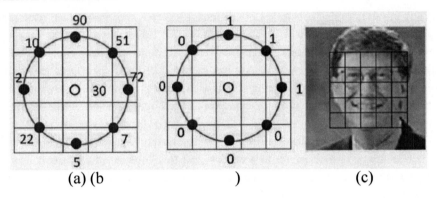

Figure 5.2: LBP operator. See details in text.

from experiments (Ojala et al., 2002), uniform patterns appear much more frequently than non-uniform patterns. Using uniform patterns is more robust to noise and can improve the efficiency of computation since the size of the codebook is significantly reduced. In some sense, these uniform patterns characterize edges with particular directions, length and scales. The operator of uniform patterns is noted as $\phi^{u2}_{P,R}$, where P is the number of pixels sampled along the circle, and R is the radius of the circle.

Ahonen et al. (2004) divided the face region into local regions as shown in Figure 5.2 (c) and used the histograms of uniform patterns inside local regions as features for face recognition. This approach significantly outperformed many popular face recognition approaches. LBP has several advantages for face recognition. First, it has high discrimination power by characterizing a very large of edges. If there are P pixels uniformly sampled along the circle, there will be $3P(P-1)$ uniform patterns. Also, when choosing different P and radius R, different uniform patterns are computed. Second, LBP is more robust to lighting variations. Local binary patterns do not change if the values of centered pixels and their neighborhoods are under the same monotonous transform functions. Third, since histograms are used as features, they are more robust to misalignment and pose variations.

However, simply adding the distances of local histograms as in Ahonen et al. (2004) includes all the uniform patterns, some of which are redundant or may deteriorate the accuracy. The efficiency is also low. Therefore, we proposed to use boosting to select a small set of uniform patterns best for face verification as described below.

Given a set of positive images $\mathcal{S}^+ = \{x_1^+, \cdots, x_{N_+}^+\}$ for a particular celebrity $\mathcal{C}$, and a set of negative images $\mathcal{S}^- = \{x_1^-, \cdots, x_{N_-}^-\}$ of other celebrities excluding $\mathcal{C}$, the goal is learn a similarity measure $F^T(x_1, x_2)$, which is high if both x_1 and x_2 belong to $\mathcal{C}$, and low, otherwise. Note in Chapter 4, we discussed detector adaptation with similarity labels, which is also about learning the relationship between two examples. However, in Chapter 4, the learned boosting classifier $F^T(x)$

still operates on a single example, while here the classifier will take two input examples. For face verification purpose, the form $F^T(x_1, x_2)$ appears more consistent with the problem itself and is hence widely used in the literature (Guo et al., 2001; Zhang et al., 2004).

From $\mathcal{S}^+$ and $\mathcal{S}^-$, a positive training set $\{(x_i^+, x_j^+), x_i^+ \in \mathcal{S}^+, x_j^+ \in \mathcal{S}^+\}$ and a negative training set $\{(x_i^+, x_j^-), x_i^+ \in \mathcal{S}^+, x_j^- \in \mathcal{S}^-\}$ are built. The similarity between an unseen test image x and $\mathcal{S}^+$, computed as:

$$\max_{x_i^+ \in \mathcal{S}^+} F^T(x, x_i^+), \tag{5.1}$$

is used as a measure to verify whether x belongs to celebrity $\mathcal{C}$ or not.

We prepared a large feature pool based on LBP as $\Phi = \{h_1, \ldots, h_L\}$, where $h_l = \phi_{P,R}^{u2}(E, k)$ is the count of the kth bin of the histogram of uniform patterns inside local region E. The features in Φ are computed using different LBP operators by choosing different P and R values and using different sizes of local regions. To compute the distance between the features of two examples, we defined the distance of features as:

$$d(h(x_1), h(x_2)) = \frac{[h(x_1) - h(x_2)]^2}{h(x_1) + h(x_2)}. \tag{5.2}$$

The overall AdaBoost-based learning algorithm is shown in Figure 5.3.

5.1.3   BOOSTED MULTI-TASK LEARNING

As mentioned earlier, we assume that for each celebrity, a small number of training examples are available for learning. If individual classifiers are learned for each celebrity, overfitting is inevitable. An alternative approach is to train a generic classifier which determines whether any two examples are from the same person or not. Such a classifier would not be person-specific, and can be used to verify any person. Many approaches such as Bayesianface (Moghaddam et al., 2000) and AdaBoost face recognition in (Yang et al., 2004; Zhang et al., 2004) used this scheme. In certain scenarios, this approach can effectively reduce the chance of overfitting since the positive and negative training sets can be very large. However, since only a single classifier is built to recognize all the faces, the recognition performance is usually not satisfactory, as shown in our experiments in Section 5.1.4. In the following, we present a novel algorithm we call Boosted Multi-Task Learning (MTL) to solve these problems.

Multi-task learning (Caruana, 1997) is a machine learning approach that learns a problem together with other related problems at the same time using a shared representation. It often leads to a better model for a single task than learning it independently, because it allows the learner to use the commonality among tasks. In our case, the tasks are the verification of multiple celebrities. Assuming there are M celebrities to be verified. A celebrity m has N_{+m} training examples $\mathcal{S}_m^+ = \{x_{m1}^+, \cdots, x_{mN_{+m}}^+\}$. There is a common negative set $\mathcal{S}^- = \{x_1^-, \cdots, x_{N_-}^-\}$, which includes training examples of other people excluding these M celebrities. For each celebrity m, a training set $\{(x_{mn}, z_{mn})\}$ is built, where $x_{mn} = (x_{mn1}, x_{mn2})$ is a pair of image examples, $z_{mn} = 1$ if both

Input

- Positive training set $\{(x_i^+, x_j^+), x_i^+ \in \mathcal{S}^+, x_j^+ \in \mathcal{S}^+\}$ and negative training set $\{(x_i^+, x_j^-), x_i^+ \in \mathcal{S}^+, x_j^- \in \mathcal{S}^-\}$.
- Feature pool $\Phi = \{h_1, \ldots, h_L\}$.
- T is the total number of weak classifiers, which can be set through cross-validation.

Initialize

- Assign initial scores for example pairs $F^0(\cdot, \cdot) = \frac{1}{2} \ln\left(\frac{N_+ - 1}{2N_-}\right)$. Note the total number of positive example pairs is $\frac{N_+(N_+ - 1)}{2}$, the total number of negative example pairs is $N_+ N_-$.

Adaboost Learning
For $t = 1, \cdots, T$:

1. For each feature h_l in the pool Φ, find the optimal threshold H for the distance function Eq. (5.2) and confidence score c_1 and c_2 to minimize the Z score L^t (1.12).
2. Select the best feature with the minimum L^t as $f_t(\cdot, \cdot)$.
3. Update $F^t(x_i, x_j) = F^{t-1}(x_i, x_j) + f_t(x_i, x_j)$,
4. Update weights of all example pairs.

Output Final classifier $F^T(\cdot, \cdot)$.

Figure 5.3: AdaBoost learning with LBP features.

x_{mn1} and x_{mn2} are in $\mathcal{S}_m^+$, and $z_{mn} = 0$ if $x_{mn1} \in \mathcal{S}_m^+$ and $x_{mn2} \in \mathcal{S}^-$. If necessary, we can choose $x_{mn2} \in \bigcup_{l \neq m} \mathcal{S}_l^+ \bigcup \mathcal{S}^-$ to expand the training set.

We use a graphical model to represent the structure of Boosted MTL, as shown in Figure 5.4. In our model, there are K boosting classifiers $\{F_k^T\}, k = 1, \cdots, K$ to be learned, where $K \leq M$. The boosting classifiers are in the following form:

$$F_k^T(x_{mn}) = \sum_{t=1}^{T} \lambda_k^t f_k^t(x_{mn}), \tag{5.3}$$

which is the score of example pair x_{mn}. And the corresponding probability is:

$$p_k(x_{mn}) = \frac{1}{1 + \exp(-F_k^T(x_{mn}))}. \tag{5.4}$$

To introduce the graphical model, let η be a multinomial hyperparameter. For a given celebrity m, the model samples a boosting classifier indexed as $c_m \in \{1, \cdots, K\}$ based on the conditional probability

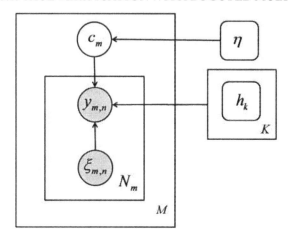

Figure 5.4: The graphical model of Boosted Multi-Task Learning.

$p(c_m|\eta)$, and uses $F_{c_m}^T$ to predict z_{mn} given x_{mn}. The joint distribution of all the training data is

$$p(\{z_{mn}\}|\{x_{mn}\}, \{F_k^T\}, \eta) = \prod_m \prod_n \left\{ \sum_{c_m} p(z_{mn}|x_{mn}, F_{c_m}^T) p(c_m|\eta) \right\} \tag{5.5}$$

where

$$p(z_{mn}|x_{mn}, F_{c_m}^T) = p_k(x_{mn})^{z_{mn}} (1 - p_k(x_{mn}))^{(1-z_{mn})}. \tag{5.6}$$

During training, an EM algorithm is adopted to learn $\{F_k^T\}$ and η.

E-step:

$$\begin{aligned}
q_{m,k}^t &\triangleq p(c_m = k|\{z_{mn}\}, \{x_{mn}\}, \{F_k^t\}, \eta^t) \\
&= \frac{\eta_k^t \prod_n p(z_{mn}|x_{mn}, F_k^t)}{\sum_{k'=1}^K \eta_{k'}^t \prod_n p(z_{mn}|x_{mn}, F_{k'}^t)}.
\end{aligned} \tag{5.7}$$

M-step:

$$\eta_k^{t+1} \propto \sum_m q_{m,k}^t. \tag{5.8}$$

and select the next weak classifier as

$$\arg\max_{f_k} \sum_{m,n} q_{m,k}^t \log\left[p(z_{mn}|x_{mn}, F_k \cup f_k)\right], \tag{5.9}$$

which can be solved by the AnyBoost approach (Mason et al., 2000). That is, let the likelihood be:

$$\begin{aligned}
L_k^T &= \sum_{m,n} q_{m,k}^T \log\left[p(z_{mn}|x_{mn}, F_k \cup f_k)\right] \\
&= \sum_{m,n} q_{m,k}^T \left[z_{mn} \log p_k(x_{mn}) + (1 - z_{mn}) \log(1 - p_k(x_{mn}))\right].
\end{aligned} \tag{5.10}$$

The weight on each example is given as the derivative of the likelihood function with respect to a change in the score of the example.

$$w_{k,mn} = \frac{\partial L_k^T}{\partial F_k^T(x_{mn})} = \sum_{m,n} q_{m,k}^T(z_{mn} - p_k(x_{mn})). \tag{5.11}$$

And the rest is similar to the standard AnyBoost procedure.

After $\{F_k^T\}$ and η have been learnt by EM, for any new input image pair x, the classifier of celebrity m outputs:

$$
\begin{aligned}
& p(z|x, \{(x_{mn}, z_{mn})\}, \{F_k^T\}, \eta) \\
&= \sum_{k=1}^{K} p(z|x, F_k^T) p(F_k^T | \{(x_{mn}, z_{mn})\}, \{F_k^T\}, \eta) \\
&= \sum_{k=1}^{K} q_{m,k}^T p(z|x, F_k^T). \tag{5.12}
\end{aligned}
$$

The algorithm is summarized in Figure 5.5.

In Boosted MTL, celebrities that have commonality in feature selection are clustered and share training data. The posterior $q_{m,k}^T$ indicates how well a boosting classifier F_k^T fits the training data of celebrity m. From Eq. (5.10) and (5.11), if F_k^T cannot explain the training data of celebrity m well, the training data of m has less contribution to the learning of F_k^T since the weight of each example is multiplied by $q_{m,k}^T$. Instead of training M boosting classifiers, in our approach only K boosting classifiers are learnt, so it is less likely to overfit. On the other hand, as shown in Eq. (5.12), the training data of each celebrity can be well explained by properly linearly combining the K boosting classifiers. Instead of requiring all the celebrities to share training data as in training a single generic boosting classifier, in boosted MTL, a set of celebrities share training data only when their training data can be well explained by the same boosting classifier. If K is smaller, the trained classifiers are less likely to overfit. We can choose the smallest K that leads to the accuracy on the training data above an expected threshold. Thus boosted MTL provides a way to maximize the generalization ability while guaranteeing certain accuracy on the training data.

5.1.4   EXPERIMENTAL RESULTS

We first compare the proposed LBP bin features with direct LBP-based recognition (Ahonen et al., 2004) and LBP-histogram-feature-based AdaBoost (Zhang et al., 2004). Following the notations in Section 5.1.2, the positive training data set S^+ has 73 images of George Bush from our database; the negative training set S^- has 8861 images of other celebrities from the LFW database (Huang et al., 2007c); the positive testing set has 523 images of George Bush from the LFW database, and the negative testing set has 4000 images of other people from the LFW database. After running a face detector (see (Zhang and Viola, 2007) and Chapter 2), face regions are normalized to 50×50 pixels.

Input

- Training set for M celebrities, $\{(x_{mn}, z_{mn})\}$, $m = 1, \cdots, M$.
- Feature pool $\Phi = \{h_1, \ldots, h_L\}$.
- Number of components K, and the total number of weak classifiers T.

Initialize

- Randomly assign a value from 1 to K to c_m, $q_{m,k} = 1$ if $c_m = k$, and 0, otherwise.
- Assign initial weights. If $c_m = k$, $w_{k,mn} = \frac{1}{N_m^+}$ if $z_{mn} = 1$ and $w_{k,mn} = -\frac{1}{N_m^-}$, otherwise. N_m^+ and N_m^- are the total number of positive and negative training examples for celebrity m, respectively. If $c_m \neq k$, $w_{k,mn} = 0$.

Boosted Multi-task Learning

For $t = 1, \cdots, T$:

1. Normalize weights such that $\sum_{mn} |w_{k,mn}| = 1$, $k = 1, \cdots, K$.
2. AnyBoost-based feature selection and vote computation following the weight computation as Eq. (5.11).
3. Update η_k using Eq. (5.8),
4. Update $q_{m,k}$ using Eq. (5.7).
5. Update $w_{k,mn}$ using Eq. (5.11).

Output Final classifier $F^T(\cdot)$. The classifier of celebrity m is given by Eq. (5.12).

Figure 5.5: Boosted multi-task learning.

Table 5.1: The true positive rates of face identification on an image set of George Bush when the false alarm rate is fixed at 0.1. The abbreviations of approaches are the same as in Figure 5.6.

LBP	boost_hist	boost_bin_38150	boost_bin_4340	boost_bin_13625
0.3910	0.5086	0.8319	0.7610	0.7055

Features are computed using three types of LBP operators ($\phi_{P=8,R=2}^{u2}$, $\phi_{P=16,R=3}^{u2}$, and $\phi_{P=16,R=4}^{u2}$), and four different sizes of local regions (10×10, 15×15, 20×20, 25×25). Local regions of the same size may have overlap. There are in total 38, 150 candidate features, among which 150 features are selected by the AdaBoost algorithm.

The ROC curves are shown in Figure 5.6. Table 5.1 shows the true positive rates when the false alarm rate is fixed as 0.1, which is an intersection of Figure 5.6. Our LBP-bin-feature-based AdaBoost approach significantly outperforms the approach that compares the distances of local histograms of LBP as described in (Ahonen et al., 2004). It also outperforms the similarly trained

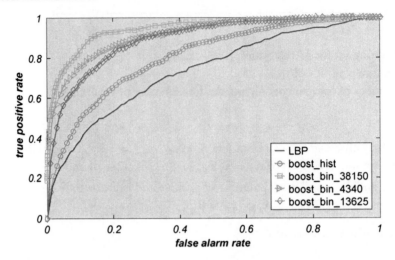

Figure 5.6: ROC curves of verifying George Bush. *LBP*: directly comparing local histograms of LBP as in Ahonen et al. (2004). *Boost_bin_38150*: AdaBoosting LBP as described in Section 5.1.2. The counts of individual bins are used as candidate features. *Boost_hist*: AdaBoosting using local histograms as features as in Zhang et al. (2004). All these three approaches above include three types of LBP operators, four difference sizes of local regions, and in total 38, 150 bins of all the local histograms. *Boost_bin_*4340 and *Boost_bin_*13625 are also AdaBoosting LBP and use counts of individual bins as features. However, *Boost_bin_*4340 uses only one type of LBP operator ($LBP^{u2}_{P=8, R=2}$) but four different sizes of local regions. *Boost_bin_*13625 uses three types of LBP operators but only one size of local regions (10×10).

AdaBoost classifier based on LBP histograms of local regions (Zhang et al., 2004). As explained in Section 5.1.2, local binary patterns characterize edges with different orientation, length and scale. The approach in (Zhang et al., 2004) kept all the edges and only selected local regions best for face verification while our approach selects both edges and regions. Thus our feature pool is much larger and the weak classifiers built from our features are simpler. Experimental results show that when using local histograms as features AdaBoosting can marginally improve the performance compared with directly using LBP, but it is much worse than using counts of individual bins as features. We also compare the performance when features are computed only using a single type of *L B P* operator (curve *Boost_bin_*4340) or only a fixed local region size (curve *Boost_bin_*13625). It shows that the performance is better when both different types of LBP operators and different local region sizes are used.

We next compare the performance of Boosted MTL with two traditionally approaches: training boost classifiers for each individual separately and training a generic boosting classifier for all the celebrities. A total of 101 celebrities are selected for this experiment. The training set has 10 examples of each celebrities from our database and 8, 861 examples of other people from the LFW

Table 5.2: The true positive rates of face identification from images of 101 celebrities using three different boosting algorithms when the false alarm rate is fixed at 0.2.

	LBP	Individual_Boost	Generic_Boost	Boosted MTL
Training	N/A	1.000	0.8067	1.000
Testing	0.3465	0.5174	0.5150	0.6098

database (Huang et al., 2007c). The testing set has 50 different examples of each celebrity from our database and 4, 000 examples of other people from the LFW database. Some examples are shown in Figure 5.17(c). This data set is very challenging since faces of celebrities have very large variations caused by factors such as makeup and aging. Figure 5.7 and Table 5.2 shows the performance of four different approaches on both the training data and the testing data: (1) directly using LBP features for verification (*LBP* as in (Ahonen et al., 2004)), (2) training a different boosting classifier for each celebrity separately (*Individual_Boost*), (3) training a generic boosting classifier to recognize all the celebrities (*Generic_Boost*), and (4) Boosted MTL with $K = 7$. *Individual_Boosting* explains the the training data perfectly but performs poorly on the testing data because of overfitting. *Generic_Boost* cannot explain the training data well and its performance is also low on the testing data. Boosted MTL can explain the training data very well and also has better performance than the other three methods on the testing data.

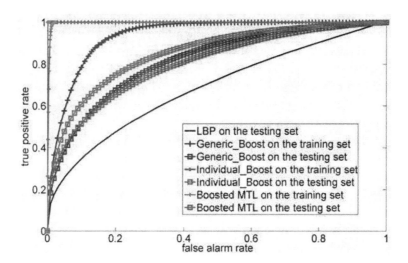

Figure 5.7: ROC curves of verifying images of 101 celebrities.

Figure 5.8: RoundTable and its captured images. (a) The RoundTable device. (b) Captured images.

5.2 BOOSTING-BASED MULTIMODAL SPEAKER DETECTION

5.2.1 INTRODUCTION

As globalization continues to spread throughout the world economy, it is increasingly common to find projects where team members are geographically distributed or even reside in different time zones. To provide a means for distributed groups to work together on shared problems, there has been an increasing interest in building special purpose devices and even "smart rooms" to support distributed meetings (Busso et al., 2005; Cutler et al., 2002; Kapralos et al., 2002; Yoshimi and Pingali, 2002). These devices often contain multiple microphones and cameras. An example device called RoundTable is shown in Fig. 5.8(a). It has a six-element circular microphone array at the base and five video cameras at the top. The captured videos are stitched into a 360 degree panorama, which gives a global view of the meeting room. The RoundTable device enables remote group members to hear and view the meeting live online. In addition, the meetings can be recorded and archived, allowing people to browse them afterward.

One of the most desired features in such distributed meeting systems is to provide remote users with a close-up of the current speaker which automatically tracks as a new participant begins to speak (Cutler et al., 2002; Kapralos et al., 2002; Yoshimi and Pingali, 2002). The speaker detection problem, however, is non-trivial. In Section 3.1.3 and Section 4.5, we have discussed the application of MILBoost and detector adaptation to improve the person detector for the panoramic images captured by the RoundTable device. For speaker detection, however, additional work is needed to identify the person who is speaking.

In existing distributed meeting systems, the two most popular speaker detection approaches are through sound source localization (SSL) (Wang and Chu, 1997; Rui et al., 2005) and SSL combined with face detection using decision level fusion (DLF) (Kapralos et al., 2002; Yoshimi and Pingali, 2002). However, they both have difficulties in practice. The success of SSL heavily depends on the levels of reverberation noise (e.g., a wall or whiteboard can act as an acoustic mirror) and ambient noise (e.g., computer fans), which are often high in many meeting rooms. Decision level fusion may help improve the final detection performance. However, low resolution face detection remains a very challenging task despite the improvement we have made in Chapter 3 and Chapter 4.

In this section, we present a novel boosting-based multimodal speaker detection (BMSD) algorithm, which integrates audio and visual multimodal information into a single boosting framework at the feature level. Specifically, we use the output of an SSL algorithm to compute audio related features for windows in the video frame. These features are then placed in the same pool as the appearance and motion visual features computed on the gray scale video frames and selected by the boosting algorithm automatically. Such a scheme is able to explicitly learn the difference between speakers and non-speakers. The proposed algorithm reduces the error rate of the SSL-only solution by 24.6% in our experiments, and the SSL and person detection DLF approach by 20.9%. In addition, it is super-efficient and achieves the above performance with merely 60 SSL and Haar basis image features. Lastly, BMSD does not require high frame rate video analysis or tight AV synchronization, which is ideal for our application. The proposed BMSD algorithm has been integrated with the RoundTable device and shipped to thousands of customers in the summer of 2007.

5.2.2   RELATED WORKS

Audio visual information fusion has been a popular approach for many research topics including speech recognition (Adjoudani and Benoît, 1996; Dupont and Luettin, 2000), video segmentation and retrieval (Hsu et al., 2004), event detection (Brand et al., 1997; Naphade et al., 2001), speaker change detection (G.Iyengar and C.Neti, 2000), speaker detection (Besson and Kunt, 2005; Fisher III et al., 2000; Hershey and Movellan, 2000; Pavlović et al., 2001) and tracking (Zotkin et al., 2002; Vermaak et al., 2001), etc. In the following paragraphs, we describe briefly a few approaches that are closely related to this section.

Audio visual synchrony is one of the most popular mechanisms to perform speaker detection. Explicitly or implicitly, many approaches measure the mutual information between audio visual signals and search for regions of high correlation and tag them as likely to contain the speaker. Representative works include Hershey and Movellan (2000), Nock et al. (2003), Besson and Kunt (2005), and Fisher III et al. (2000). Cutler and Davis (2000) instead learned the audio visual correlation using a time-delayed neural network (TDNN). Approaches in this category often need just a single microphone, and they rely on the synchrony only to identify the speaker. Most of them require a good frontal face to work well.

Another popular approach is to build graphical models for the observed audio visual data and infer the speaker location probabilistically. Pavlović et al. (2001) proposed to use dynamic Bayesian networks (DBN) to combine multiple sensors/detectors and decide whether a speaker is present in front of a smart kiosk. Beal et al. (2002) built a probabilistic generative model to describe the observed data directly using an EM algorithm and estimated the object location through Bayesian inference. Brand et al. (1997) used coupled hidden Markov models to model the relationship between audio visual signals and classify human gestures. Graphical models are a natural way to solve multimodal problems and are often intuitive to construct. However, their inference stage can be time-consuming and would not fit into our tight computation budget.

Audio visual fusion has also been applied for speaker tracking, in particular, those based on particle filtering (Zotkin et al., 2002; Vermaak et al., 2001; Chen and Rui, 2004; Nickel et al., 2005). In the measurement stage, audio likelihood and video likelihood are both computed for each sample to derive its new weight. It is possible to use these likelihoods as measures for speaker detection, though such an approach can be very expensive if all the possible candidates in the frame need to be scanned.

In real-world applications, the two most popular speaker detection approaches are still SSL-only and SSL combined with face detection for decision level fusion (Cutler et al., 2002; Kapralos et al., 2002; Yoshimi and Pingali, 2002). For instance, the iPower 900 teleconferencing system from Polycom uses an SSL-only solution for speaker detection (Wang and Chu, 1997). Kapralos et al. (2002) used a skin-color-based face detector to find all the potential faces and detect speech along the directions of these faces. Yoshimi and Pingali (2002) took the audio localization results and used a face detector to search for nearby faces in the image. Busso et al. (2005) adopted Gaussian mixture models to model the speaker locations and fused the audio and visual results probabilistically with temporal filtering.

As mentioned earlier, speaker detection based on SSL-only is sensitive to reverberation and ambient noises. The DLF approach, on the other hand, has two major drawbacks in speaker detection. First, when SSL and face detection operate separately, the correlation between audio and video, either at high frame rate or low frame rate, is lost. Second, a full-fledged face detector can be unnecessarily slow because many regions in the video can be skipped if their SSL confidence is too low. Limiting the search range of face detection near SSL peaks, however, is difficult because it is hard to find a universal SSL threshold for all conference rooms. Moreover, this can introduce bias towards the decision made by SSL. The proposed algorithm uses a boosted classifier to perform feature level fusion of information in order to minimize computation time and maximize robustness. We will show the superior performance of BMSD by comparing it with the SSL-only and DLF approaches in Section 5.2.7.

5.2.3 SOUND SOURCE LOCALIZATION

Because the panoramic video available for speaker detection is at very low resolution, it is very challenging even for humans to tell who is the active speaker. Consequently, we first investigated the idea of building a better SSL algorithm that is less sensitive to reverberation or ambient noises. We developed a novel maximum-likelihood(ML)-based sound source localization algorithm that is both efficient and robust to reverberation and noises (Zhang et al., 2007a). In the proposed BMSD algorithm, audio related features are extracted from the output of the ML-based SSL algorithm instead of the original audio signal. For the completeness of this chapter, we provide a brief review of the ML-based SSL algorithm in this section, and we refer the readers to (Zhang et al., 2007a) for more details.

Consider an array of P microphones (In the case of RoundTable, there are a total of 6 *directional* microphones on the base of the device). Given a source signal $s(t)$, the signals received at these microphones can be modeled as (Gustafsson et al., 2001):

$$x_i(t) = \alpha_i s(t - \tau_i) + h_i(t) \otimes s(t) + n_i(t), \qquad (5.13)$$

where $i = 1, \cdots, P$ is the index of the microphones, τ_i is the time of propagation from the source location to the ith microphone; α_i is a gain factor that includes the propagation energy decay of the signal, the gain of the corresponding microphone, the directionality of the source and the microphone, etc.; $n_i(t)$ is the noise sensed by the ith microphone; $h_i(t) \otimes s(t)$ represents the convolution between the environmental response function $h_i(t)$ and the source signal, often referred as the *reverberation*.

In the frequency domain, the equivalent form of the above model is:

$$X_i(\omega) = \alpha_i(\omega)S(\omega)e^{-j\omega\tau_i} + H_i(\omega)S(\omega) + N_i(\omega), \qquad (5.14)$$

where we allow the α_i to vary with frequency. In a vector form:

$$\mathbf{X}(\omega) = S(\omega)\mathbf{G}(\omega) + S(\omega)\mathbf{H}(\omega) + \mathbf{N}(\omega), \qquad (5.15)$$

where

$$
\begin{aligned}
\mathbf{X}(\omega) &= [X_1(\omega), \cdots, X_P(\omega)]^T, \\
\mathbf{G}(\omega) &= [\alpha_1(\omega)e^{-j\omega\tau_1}, \cdots, \alpha_P(\omega)e^{-j\omega\tau_P}]^T, \\
\mathbf{H}(\omega) &= [H_1(\omega), \cdots, H_P(\omega)]^T, \\
\mathbf{N}(\omega) &= [N_1(\omega), \cdots, N_P(\omega)]^T.
\end{aligned}
$$

Our ML-based SSL algorithm makes the assumption that the combined total noise,

$$\mathbf{N}^c(\omega) = S(\omega)\mathbf{H}(\omega) + \mathbf{N}(\omega), \qquad (5.16)$$

follows a zero-mean, independent between frequencies, joint Gaussian distribution, i.e.,

$$p(\mathbf{N}^c(\omega)) = \rho \exp\left\{ -\frac{1}{2}[\mathbf{N}^c(\omega)]^H \mathbf{Q}^{-1}(\omega)\mathbf{N}^c(\omega) \right\}, \qquad (5.17)$$

where ρ is a constant; superscript H represents Hermitian transpose, $\mathbf{Q}(\omega)$ is the covariance matrix of the combined noise and can be estimated from the received audio signals.

The likelihood of the received signals can be written as:

$$p(\mathbf{X}|S, \mathbf{G}, \mathbf{Q}) = \prod_\omega p(\mathbf{X}(\omega)|S(\omega), \mathbf{G}(\omega), \mathbf{Q}(\omega)), \qquad (5.18)$$

where

$$p(\mathbf{X}(\omega)|S(\omega), \mathbf{G}(\omega), \mathbf{Q}(\omega)) = \rho \exp\left\{ -J(\omega)/2 \right\}, \qquad (5.19)$$

$$J(\omega) = [\mathbf{X}(\omega) - S(\omega)\mathbf{G}(\omega)]^H \mathbf{Q}^{-1}(\omega)[\mathbf{X}(\omega) - S(\omega)\mathbf{G}(\omega)]. \tag{5.20}$$

The goal of the proposed sound source localization is thus to maximize the above likelihood, given the observations $\mathbf{X}(\omega)$, gain matrix $\mathbf{G}(\omega)$ and noise covariance matrix $\mathbf{Q}(\omega)$. Note the gain matrix $\mathbf{G}(\omega)$ requires information about where the sound source comes from; hence, the optimization is usually solved through hypothesis testing. That is, hypotheses are made about the source source location, which can be used to compute the corresponding $\mathbf{G}(\omega)$. The likelihood is then measured. The hypothesis that results in the highest likelihood is determined to be the output of the SSL algorithm.

We showed that the solution to such a ML problem is to use hypothesis testing to measure the maximum value of

$$L = \int_{\omega} \frac{[\mathbf{G}^H(\omega)\mathbf{Q}^{-1}(\omega)\mathbf{X}(\omega)]^H \mathbf{G}^H(\omega)\mathbf{Q}^{-1}(\omega)\mathbf{X}(\omega)}{\mathbf{G}^H(\omega)\mathbf{Q}^{-1}(\omega)\mathbf{G}(\omega)} d\omega. \tag{5.21}$$

Under certain simplified conditions, the above criterion can be computed efficiently. More specifically, one may hypothesize the source locations, measure and maximize the following objective function:

$$J = \int_{\omega} \frac{\left| \sum_{i=1}^{P} \frac{X_i(\omega)e^{j\omega\tau_i}}{\kappa_i(\omega)} \sqrt{|X_i(\omega)|^2 - E\{|N_i(\omega)|^2\}} \right|^2}{\sum_{i=1}^{P} \frac{1}{\kappa_i(\omega)}(|X_i(\omega)|^2 - E\{|N_i(\omega)|^2\})} d\omega, \tag{5.22}$$

where

$$\kappa_i(\omega) = \gamma |X_i(\omega)|^2 + (1 - \gamma)E\{|N_i(\omega)|^2\}, \tag{5.23}$$

and γ is a parameter modeling the severity of room reverberation. Interested readers are referred to (Zhang et al., 2007a) for more details about the assumptions and derivations.

The ML-based SSL algorithm significantly improved the localization accuracy over traditional SSL algorithms such as SRP-PHAT (Brandstein and Silverman, 1997), in particular under very noisy environments. However, it may still be deceived by heavy reverberation and points to directions where no people are there. In the next section, we present the boosting-based multimodal speaker detection algorithm that aims to use visual information to further improve the speaker detection performance.

5.2.4 BOOSTING-BASED MULTIMODAL SPEAKER DETECTION

Our speaker detection algorithm adopts the same boosting algorithm as in Fig. 2.2 to learn the difference between speakers and non-speakers. It computes both audio and visual features and places them in a common feature pool for the boosting algorithm to select. This has a number of advantages. First, the boosting algorithm explicitly learns the difference between a speaker and a non-speaker, thus it targets the speaker detection problem more directly. Second, the final classifier can contain both audio and visual features, which implicitly explores the correlation between the audio and visual information if they coexist after the feature selection. Third, thanks to the cascade structure, audio features selected early in the learning process will help eliminate many non-speaker windows, which greatly improves the detection speed. Finally, since all the audio visual features are in the same pool, there is no bias toward either modality.

We use the same visual features introduced in Section 3.1.3. For audio features, we extract them based on the output of the hypothesis testing process during SSL. Note the microphone array on the RoundTable has a circular geometry. Hence the SSL can only provide reliable 1D azimuth of the sound source location through hypothesis testing. We obtain a 1D array of numbers between 0 and 1 following Eq. (5.21), denoted as $L_a(\theta), \theta = 0, \alpha, \cdots, 360 - \alpha$. The hypothesis testing is done for every α degrees. In the current implementation, $\alpha = 4$ gives good results. We perform SSL at 1 frame per second (FPS). The synchronization between audio and video is guaranteed to be within 100 milliseconds. For computing audio features for detection windows in the video frames, we map $L_a(\theta)$ to the image coordinate as:

$$L_a(u) = L_a\big(\theta(u)\big), u = 1, 2, \cdots, U, \tag{5.24}$$

where U is the width of the panoramic images, and $\theta(u)$ is the mapping function.

Figure 5.9: Compute SSL features for BMSD. (a) Original image. (b) SSL image. Bright intensity represents high likelihood. Note the peak of the SSL image does not correspond to the actual speaker (the right-most person), indicating a failure for the SSL-only solution.

It is not immediately clear what kind of audio features can be computed for each detection window from the above 1D likelihood array. One possibility is to create a 2D image out of the 1D array by duplicating the values along the vertical axis, as shown in Fig. 5.9(b) (an similar approach was taken in (Goodridge, 1997)). One can treat this image the same as the other ones and compute rectangle features on this image. However, the local variation of SSL is a very poor indicator of the speaker location. We instead compute a set of audio features for each detection window with respect to the whole SSL likelihood distribution. The global maximum, minimum and average SSL outputs

Table 5.3: Audio Features extracted from the SSL likelihood function. Note the 15th feature is a binary one which tests if the local region contains the global peak of SSL.

1. $\dfrac{L^l_{max}-L^g_{min}}{L^g_{max}-L^g_{min}}$	2. $\dfrac{L^l_{min}-L^g_{min}}{L^g_{max}-L^g_{min}}$	3. $\dfrac{L^l_{avg}-L^g_{min}}{L^g_{max}-L^g_{min}}$	4. $\dfrac{L^l_{mid}-L^g_{min}}{L^g_{max}-L^g_{min}}$	5. $\dfrac{L^l_{max}}{L^l_{min}}$
6. $\dfrac{L^l_{max}}{L^l_{avg}}$	7. $\dfrac{L^l_{min}}{L^l_{avg}}$	8. $\dfrac{L^l_{mid}}{L^l_{avg}}$	9. $\dfrac{L^l_{max}-L^l_{min}}{L^l_{avg}}$	10. $\dfrac{L^l_{max}}{L^g_{max}}$
11. $\dfrac{L^l_{min}}{L^g_{max}}$	12. $\dfrac{L^l_{avg}}{L^g_{max}}$	13. $\dfrac{L^l_{mid}}{L^g_{max}}$	14. $\dfrac{L^l_{max}-L^l_{min}}{L^g_{max}}$	15. $L^g_{max} - L^l_{max} < \epsilon$

are first computed respectively as

$$\text{global maximum:} \quad L^g_{max} = \max_u L_a(u),$$

$$\text{global minimum:} \quad L^g_{min} = \min_u L_a(u),$$

$$\text{global average:} \quad L^g_{avg} = \frac{1}{U} \sum_u L_a(u). \tag{5.25}$$

Let the left and right boundaries of a detection window be u_0 and u_1. Four local values are computed as follows:

$$\text{local maximum:} \quad L^l_{max} = \max_{u_0 \le u \le u_1} L_a(u),$$

$$\text{local minimum:} \quad L^l_{min} = \min_{u_0 \le u \le u_1} L_a(u),$$

$$\text{local average:} \quad L^l_{avg} = \frac{1}{u_1 - u_0} \sum_{u_0 \le u \le u_1} L_a(u),$$

$$\text{middle output:} \quad L^l_{mid} = L_a\left(\frac{u_0 + u_1}{2}\right). \tag{5.26}$$

We then extract 15 features out of the above values, as shown in Table 5.3.

It is important to note that the audio features used here have no discrimination power along the vertical axis. Nevertheless, across different columns, the audio features can vary significantly; hence, they can still be very good weak classifiers. We let the boosting algorithm decide if such classifiers are helpful. From the experiments in Section 5.2.7, SSL features are among the top features selected by the boosting algorithm.

5.2.5   MERGE OF DETECTED WINDOWS

Depending on the value of the final threshold for the cascade detector, there are usually hundreds of windows that are classified as positive by the trained classifier in one video frame. This is different from the typical face detectors, where there are usually 1 to 10 positive windows around a true face. One possibility to reduce the number of detected positive windows is to increase the value of the final threshold. Unfortunately, this will also increase the false negative rate (FNR), namely, the frequency of scenarios where no window is classified as positive for a frame with people speaking.

The detection accuracy/FNR tradeoff exists in almost any machine-learning-based classifiers. Speaker detection with SSL and low resolution video inputs is a very challenging problem, and it is difficult to find a single final threshold that delivers satisfactory accuracy and FNR at the same time. Fortunately, speaker detection is also a unique problem for the following reason: the RoundTable device will only output a single speaker's high resolution video to the remote site, even when multiple people are talking simultaneously. That is, despite the hundreds of raw positive windows being detected, only a single output window is needed. This allows us to develop novel merging algorithms to help increase the accuracy of the final detection results. In the following paragraphs, we present two merging algorithms that we implemented for this purpose: projection and merge (PAM) and top N merge (TNM).

Let the positively detected windows be $R_k = \{u_0^k, u_1^k, v_0^k, v_1^k\}, k = 1, \cdots, K$, where K is the total number of positive windows, u_0 and u_1 represent the left and right boundaries of the window, and v_0 and v_1 represent the top and bottom boundaries of the window. The PAM algorithm first projects all these windows onto the horizontal axis, giving:

$$r(u) = \sum_k \delta(u_0^k \le u \le u_1^k), \qquad (5.27)$$

where $u = 1, \cdots, U$ is the index of horizontal pixel positions, $\delta(\cdot)$ is 1 if the condition is satisfied, and 0, otherwise. We then locate the peak of the function $r(u)$:

$$\hat{u} = \arg\max_u r(u). \qquad (5.28)$$

The merged output window is computed as:

$$R_{\text{out}} = \text{Avg}\{R_k | u_0^k \le \hat{u} \le u_1^k\}. \qquad (5.29)$$

That is, the merged window is the average of all positive windows that overlap with the peak of $r(u)$.

The TNM algorithm relies on the fact that the final score of each positively detected window, i.e., $s(x) = \sum_{t=1}^T \alpha_t h_t(x)$, is a good indicator of how likely the window is actually a positive window. In TNM, we first pick the N positive windows that have the highest probability $p(x)$, and then merge these top N windows with the previous PAM algorithm. The typical value of N is 5 to 9 in the currently implementation. As will be shown in Section 5.2.7, this simple approach can significantly improve the speaker detection accuracy at the same false negative rate.

5.2.6   ALTERNATIVE SPEAKER DETECTION ALGORITHMS

The most widely used approach to speaker detection is SSL (Rui et al., 2005; Wang and Chu, 1997). Given the SSL likelihood as $L_a(u), u = 1, 2, \cdots, U$, we simply look for the peak likelihood to obtain the speaker direction:

$$\hat{u} = \arg\max_u L_a(u). \qquad (5.30)$$

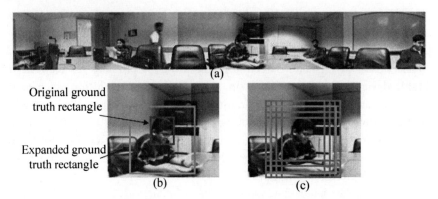

Figure 5.10: Create positive examples from the ground truth. (a) Original video frame. (b) Close-up view of the speaker. The blue rectangle is the head box; the green box is the expanded ground truth box. (c) All gray rectangles are considered positive examples.

This method is extremely simple and fast, but its performance can vary significantly across different conference rooms depending on the room acoustic properties.

The second approach is to design a multi-person detector (MPD) and fuse its results with SSL output probabilistically. In Section 3.1.3 and 4.5, we have discussed the learning and adaptation of MPD. The MPD output is a list of head boxes. To fuse with the 1D SSL output, a 1D video likelihood function can be created from these boxes through kernel methods, i.e.:

$$L_v(u) = (1 - \varepsilon) \sum_{n=1}^{N} e^{-\frac{(u-u_n)^2}{2\sigma^2}} + \varepsilon, \tag{5.31}$$

where N is the number of detected boxes; u_n is the horizontal center for the nth box; σ is $\frac{1}{3}$ of the average head box width; ε is a small constant to represent the likelihood of a person when MPD has detected nothing nearby. Assuming the audio and visual likelihoods are independent, the total likelihood is computed as:

$$L(u) = L_a^{\beta}(u) * L_v^{1-\beta}(u), \tag{5.32}$$

where β is a parameter controlling the tradeoff between SSL and MPD. We pick the highest peak in $L(u)$ as the horizontal center of the active speaker. The height and scale of the speaker is determined by its nearest detected head box.

5.2.7 EXPERIMENTAL RESULTS

Experiments were conducted on a large set of video sequences captured in real-world meetings. The training data set consists of 93 meetings (3204 labeled frames) collected in more than 10 different conference rooms. Each meeting sequence is about 4 minutes long. The labeled frames are spaced at

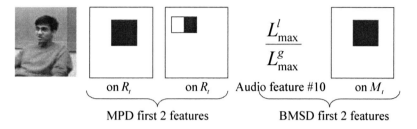

Figure 5.11: Top features for MPD and BMSD.

least 1 second apart. In each labeled frame, there are one or multiple people speaking [1]. The speaker
is marked by a hand-drawn box around the head of the person. Since the human body can provide
extra cues for speaker detection, we expand every head box with a constant ratio to include part of
upper body, as shown in Fig. 5.10(b). Rectangles that are within a certain translation and scaling
limits of the expanded ground truth boxes are used as positive examples (Fig. 5.10(c)). The remaining
rectangles in the videos are all treated as negative examples. The minimum detection window size
is 35×35, and the 93 sequences comprise over 100 million examples (including both positive and
negative examples) for training.

The test data set consists of 82 meetings (1502 labeled frames) recorded in 10 meeting rooms.
Most of the rooms are different from those used in the training data set.

In the following experiments, we will compare the performance of the SSL-only, SSL+MPD
DLF and BMSD algorithms. The BMSD classifier contains 60 audio/visual features (weak classifiers) selected by the boosting process. The MPD algorithm is trained with the same training data
set using the same AdaBoost algorithm as BMSD, except that we use all visible people as positive
examples. We allow the MPD to contain 120 weak classifiers, doubling the amount of features in
BMSD. Note in the BMSD training process, the negative examples include people in the meeting
room that were not talking. We expect BMSD to learn explicitly the difference between speakers
and non-speakers.

For a quick comparison, Fig. 5.11 shows the first two features of MPD and BMSD selected by
the boosting process. For MPD, both features are on the running difference image. The first feature
favors rectangles where there is motion around the head region. The second feature describes that
there is a motion contrast around the head region. In BMSD, the first feature is an audio feature,
which is the ratio between the local maximum likelihood and the global maximum likelihood. The
second feature is a motion feature similar to the first feature of MPD, but on the 3-frame difference

---

[1]We assume that the audio processing module (including SSL) has the full responsibility to determine whether there is anyone
speaking in the conference room. Such classification can certainly make mistakes. However, we found adding frames where no
one is speaking while the audio module claims someone speaking can only confuse the BMSD training process, leading to worse
performance on speaker detection.

image. It is obvious that although audio features do not have discrimination power along the vertical axis, they are still very good features to be selected into the BMSD classifier.

To measure if the detected rectangle is a true positive detection, we use the following criterion. Let the ground truth face be $R_g = \{u_0^g, u_1^g, v_0^g, v_1^g\}$, and the detected box be $R_{out} = \{u_0^o, u_1^o, v_0^o, v_1^o\}$. A true positive detection must satisfy:

$$\left| \frac{u_0^o + u_1^o}{2} - \frac{u_0^g + u_1^g}{2} \right| \leq \max(30, u_1^g - u_0^g), \tag{5.33}$$

where 30 is a tolerance number computed based on the resolution of the panoramic video for speaker detection (1056×144 pixels) and the resolution of the enlarged speaker view (320×240 pixels).

The current implementation of RoundTable does not perform any scale dependent zoom for the speaker; hence, the horizontal accuracy is the only criterion we measure. Note the device is actually capable of capturing very high resolution panoramic images (4000×600 pixels after stitching the 5 raw images). The proposed BMSD is accurate enough in vertical location and scale in order to enable full digital zooming, and this functionality will be included in future versions of RoundTable.

When only SSL is applied to compute the speaker location, we achieved a speaker detection rate (SDR) of 93.60%, a person detection rate (PDR) of 96.30%, and a false negative rate (FNR) of 4.50%. Here the SDR, PDR and FNR are defined, respectively, as follows:

$$SDR = \frac{N_{ms}}{N_{det}}, \tag{5.34}$$

$$PDR = \frac{N_{mp}}{N_{det}}, \tag{5.35}$$

$$FNR = 1 - \frac{N_{det}}{N_{total}} \tag{5.36}$$

where N_{total} is the total number of frames being tested; N_{det} is the number of frames where a speaker is detected; N_{ms} is the number of frames where the detected speaker match any ground truth speaker; N_{mp} is the number of frames where the detected speaker match any ground truth person.

Note the FNR of SSL-only speaker detection is non-zero. This means there are frames that contain people talking, but the audio module decided that no one is speaking. Since in the current implementation, both SSL+MPD DLF and BMSD work only when SSL decides that there exists a speaker; their FNR will always be higher or equal to 4.50% on this test data set.

The PDR is unique to our RoundTable application. It reports the accuracy of the speaker detector pointing to a person (may or may not be the speaker), instead of a wall or arbitrary objects. We found, in practice, the error of pointing to a non-speaking person is often much more tolerable than pointing to, say, a blank wall.

We trained an MPD detector based on the same training data, except that all the people in the training data set are used as positive examples. Only visual features are selected by the boosting process. Fig. 5.12 shows the performance of MPD with 120 visual features (further increasing the

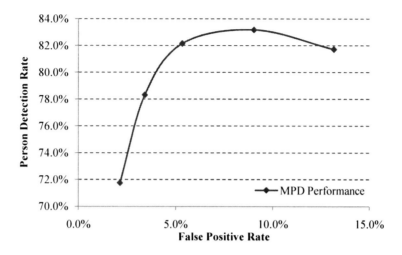

Figure 5.12: Performance of the MPD detector on the test data set.

number of features has limited improvements). The horizontal axis is false positive rate, defined as the number of false positive detections divided by the total number of ground truth person. The vertical axis is person detection rate, defined as the percentage of people that are accurately detected. Note MPD may output multiple detected person in each video frame; hence, the merging schemes discussed in Section 5.2.5 cannot be applied. We used the traditional scheme as in (Viola and Jones, 2001) for merging the rectangles.

We perform decision level fusion for the SSL and MPD as described in Section 5.2.6. From Fig. 5.12, we noted that the person detection rate of MPD is around 80%; hence, we use $\varepsilon = 0.2$ in Eq. (5.31). By varying the controlling parameter β and the MPD's thresholds, we obtained a family of performance curves, as shown in Fig. 5.13. It can be seen that the optimal β value for SSL+MPD DLF is around 0.95. The SDR reaches as high as 94.69%, at the FNR of 5.65%.

The optimal β value is high, which means that more emphasis is given to the SSL instead of the MPD. This is expected because the MPD's raw performance is relatively poor. For comparison, we also performed decision level fusion between SSL and the *ground truth* labels. Such a fusion is unrealistic, but it may provide insights on what would be the ultimate performance SSL+MPD DLF can achieve (if MPD is perfect). We obtained an SDR of 95.45% at the FNR of 4.50%.

The performance of the BMSD detector is summarized in Fig. 5.14. We used $N = 5$ for TNM merge. Note the TNM merging algorithm has significantly improved the BMSD's performance compared with PAM, in particular, when the required FNR is low. This is because when the FNR is low, one must choose to use a very low threshold for the detector. PAM merges all the windows above this threshold, which is error-prone. In contrast, TNM only picks the top N scored window to merge, which has effectively created an adaptive way to threshold the scores of the detected windows.

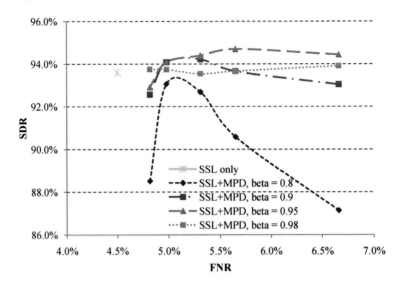

Figure 5.13: Performance of SSL+MPD decision level fusion on the test data set.

The BMSD outperforms both SSL only and SSL+MPD decision level fusion. At the FNR of 4.50%, BMSD achieves a SDR value of 95.29%. Compared with the SDR of 93.60% for the SSL only approach, we achieved a decrease of 24.6% in error. If a higher FNR is acceptable, e.g., at 5.65%, BMSD achieves a SDR value of 95.80%. Compared with the SDR value of 94.69% for SSL+MPD DLF, the improvement is a decrease of 20.9% in error. Note the BMSD is also much faster than SSL+MPD, because the BMSD detector uses only 60 features (weak classifiers), and the MPD uses 120 features.

Fig. 5.15 shows the person detection rate of the three algorithms on the test data set. As mentioned earlier, the errors that a speaker detector points to a non-speaking person is much more tolerable than pointing to a blank wall. It can be seen from Fig. 5.15 that BMSD outperforms SSL only and SSL+MPD DLF significantly in person detection rate. Compared with SSL only, at 4.50% FNR, the BMSD's PDR achieves 98.48% in contrast to 96.30% for SSL only – a decrease of 58.9% in error. Compared with SSL+MPD DLF, at 5.65% FNR, BMSD has a PDR value of 98.92% in contrast to 97.77% for SSL+MPD DLF – a decrease of 51.6% in error. For quick reference, we summarize the performance comparisons between SSL-only, SSL+MPD DLF and BMSD in Fig. 5.16.

Fig. 5.17 shows a number of examples of the detection results using BMSD. Fig. 5.1(a) and (b) are correctly detected examples, and Fig. 5.17(c) shows a typical failure example. We notice that most detection failures happen when the wall reflects the sound waves and causes confusion for the SSL. If there is a person sitting at the same place where the sound waves are reflected, it is often difficult to find the correct speaker given the low resolution and frame rate of our visual input.

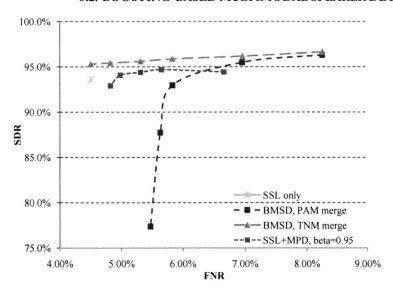

Figure 5.14: Performance of BMSD algorithm on the test data set.

Consequently, slight performance degradation may occur if there are too many people in the room. Fortunately, such a degradation should not cause the algorithm to perform worse than the SSL-only solution. In Fig. 5.1(b), BMSD found the correct speaker despite the wrong SSL decision. This is a good example showing that BMSD is also learning differences between speakers and non-speakers. For instance, the speakers tend to have more motion than non-speakers.

It is worth mentioning that an additional layer, namely a *virtual director*, is needed in order to control the switching between speakers. While the speaker detector can run up to 15 FPS, research studies have shown that switching too frequently between shots can be very distracting (Cutler et al., 2002). In the RoundTable implementation, we follow the simple rule that if the camera has just switched from one speaker to the other, it will stay there for at least 2 seconds before it can switch to another speaker. In addition, a switch is made only if a speaker has been consistently detected over a period of half a second in order to avoid spontaneous switches due to short comments.

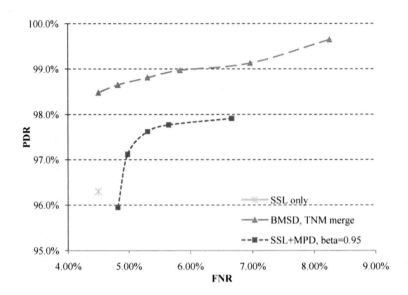

Figure 5.15: Person detection performance of various algorithms on the test data set.

		SSL- only	SSL+MPD DLF	BMSD	Decrease in error
FNR= 4.50%	SDR	93.60%	-	95.29%	24.6%
	PDR	96.30%	-	98.48%	58.9%
FNR= 4.65%	SDR	-	94.69%	95.80%	20.9%
	PDR	-	97.77%	98.92%	51.6%

Figure 5.16: Performance comparison table of various algorithms.

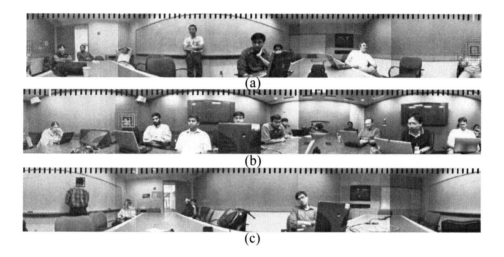

Figure 5.17: Examples of the detection results using BMSD. The bars on the top of each image show the strength of SSL likelihood, where the yellow bar is the peak of the SSL. The dark blue rectangles in the images are the ground truth active speaker, the semi-transparent rectangles are the raw detection results, and the light green rectangles are the detection results after merging.

CHAPTER 6

Conclusions and Future Work

In this book, we presented a few boosting-based face detection and adaptation technologies. They are all rooted from some well known work in the machine learning literature, such as the confidence rated boosting (Schapire and Singer, 1999), the statistical view of boosting (Friedman et al., 1998), the AnyBoost framework (Mason et al., 2000), which views boosting as a gradient decent process, and the general idea of multiple instance learning (Nowlan and Platt, 1995).

In particular, in Chapter 2, we described a boosting learning framework based on importance sampling, weight trimming and multiple instance pruning, which is capable of learning highly efficient and accurate detectors with a huge amount of training data. In Chapter 3, we discussed multiple instance learning boosting and multiple category boosting, which intends to address the problem that the object locations or subcategory labels are uncertain. In Chapter 4, we further proposed an adaptation scheme for logistic- and boosting-based classifiers, which helps a generically trained detector to adapt to a new environment. Finally, in Chapter 5, we presented two interesting applications for boosting learning, namely, face verification and speaker detection. In addition to new features applied in these two applications, the multiple task learning framework was also useful in handling scenarios where the training data set has limited sizes.

Face detection techniques have been increasingly used in real-world applications and products. For instance, most digital cameras today have built-in face detectors, which can help the camera to do better auto-focusing and auto-exposure. Digital photo management softwares such as Apple's iPhoto, Google's Picasa and Microsoft's Windows Live Photo Gallery all have excellent face detectors to help tagging and organizing people's photo collections. On the other hand, as was pointed in a recent technical report by Jain and Learned-Miller (2010), face detection in completely unconstrained settings remains a very challenging task, particularly due to the significant pose and lighting variations. In our in-house tests, the state-of-the-art face detectors can achieve about 50-70% detection rate, with about 0.5-3% of the detected faces being false positives. Consequently, we believe there is still a lot of work that can be done to further improve the performance.

The most straightforward future direction is to further improve the learning algorithm and features. The Haar features used in the work by Viola and Jones (2001) are very simple and effective for frontal face detection, but they are less ideal for faces at arbitrary poses. Complex features may increase the computational complexity, though they can be used in the form of a post-filter and still be efficient, which may significantly improve the detector's performance. Regarding learning, the boosting learning scheme is great if all the features can be pre-specified. However, other learning algorithms such as SVM or convolutional neural networks can often perform equally well, with built-in mechanisms for new feature generation.

The modern face detectors are mostly appearance-based methods, which means that they need training data to learn the classifiers. Collecting a large amount of ground truth data remains a very expensive task, which certainly demands more research. Schemes such as multiple instance learning boosting and multiple category boosting are helpful in reducing the accuracy needed for the labeled data, though ideally one would like to leverage unlabeled data to facilitate learning. Unsupervised or semi-supervised learning schemes would be very ideal to reduce the amount of work needed for data collection.

Another interesting idea to improve face detection performance is to consider the contextual information. Human faces are most likely linked with other body parts, and these other body parts can provide a strong cue of faces. There has been some recent work on context-based object categorization (Galleguillos and Belongie, 2010) and visual tracking (Yang et al., 2009). One scheme of using local context to improve face detection was also presented in (Kruppa et al., 2003), and we think that is a very promising direction to pursue.

In environments which have low variations, adaptation could bring very significant improvements to face detection. We showed the idea in Chapter 4 with a simple scheme based on the Taylor expansion of the learning target function such as logistic regression or boosting, though there are many other possibilities. Unlike in other domains such as speech recognition and handwriting recognition, where adaptation has been indispensable, adaptation for visual object detection has received relatively little attention. We strongly believe that this is a great direction for future work.

Bibliography

Y. Abramson and B. Steux. YEF* real-time object detection. In *International Workshop on Automatic Learning and Real-Time*, 2005. DOI: 10.1504/IJISTA.2007.012476 8, 12

A. Adjoudani and C. Benoît. On the integration of auditory and visual parameters in an HMM-based ASR. In *Speechreading by Humans and Machines*, pages 461–471. Springer, Berlin, 1996. 95

T. Ahonen, A. Hadid, and M. Pietikäinen. Face recognition with local binary patterns. In *Proc. of ECCV*, 2004. DOI: 10.1007/978-3-540-24670-1_36 11, 85, 86, 90, 91, 92, 93

S. Avidan. Ensemble tracking. In *CVPR*, 2005. DOI: 10.1109/CVPR.2005.144 69, 76

B. Babenko, P. Dollár, Z. Tu, and S. Belongie. Simultaneous learning and alignment: Multi-instance and multi-pose learning. In *Workshop on Faces in 'Real-Life' Images: Detection, Alignment, and Recognition*, 2008. 16, 22, 26, 56

S. Baluja, M. Sahami, and H. A. Rowley. Efficient face orientation discrimination. In *Proc. of ICIP*, 2004. DOI: 10.1109/ICIP.2004.1418823 8, 12

M. Beal, H. Attias, and N. Jojic. Audio-video sensor fusion with probabilistic graphical models. In *Proc. of ECCV*, 2002. DOI: 10.1007/3-540-47969-4_49 95

P. Besson and M. Kunt. Information theoretic optimization of audio features for multimodal speaker detection. Technical report, Signal Processing Institute, EPFL, 2005. 95

S. M. Bileschi and B. Heisele. Advances in component-based face detection. In *Pattern Recognition with Support Vector Machines Workshop*, 2002. DOI: 10.1007/3-540-45665-1_11 25

C. Bishop and P. Viola. Learning and vision: Discriminative methods. In *ICCV Course on Learning and Vision*, 2003. 4, 15, 16

C. M. Bishop. *Pattern Recognition and Machine Learning*. Springer Science+Business Media, LLC, 2006. 70, 72, 77

L. Bourdev and J. Brandt. Robust object detection via soft cascade. In *Proc. of CVPR*, 2005. DOI: 10.1109/CVPR.2005.310 16, 18, 19, 29, 30, 38, 42, 43, 44

M. Brand, N. Oliver, and A. Pentland. Coupled hidden Markov models for complex action recognition. In *Proc. of IEEE CVPR*, 1997. DOI: 10.1109/CVPR.1997.609450 95

M. Brandstein and H. Silverman. A robust method for speech signal time-delay estimation in reverberant rooms. In *Proc. of ICASSP*, 1997. DOI: 10.1109/ICASSP.1997.599651 98

S. C. Brubaker, J. Wu, J. Sun, M. D. Mullin, and J. M. Rehg. On the design of cascades of boosted ensembles for face detection. Technical report, Georgia Institute of Technology, GIT-GVU-05-28, 2005. DOI: 10.1007/s11263-007-0060-1 11, 15, 16, 18, 19, 35

C. Busso, S. Hernanz, C. Chu, S. Kwon, S. Lee, P. Georgiou, I. Cohen, and S. Narayanan. Smart room: participant and speaker localization and identification. In *Proc. of IEEE ICASSP*, 2005. DOI: 10.1109/ICASSP.2005.1415605 94, 96

M. M. Campos and G. A. Carpenter. S-tree: Self-organizing trees for data clustering and online vector quantization. *Neural Networks*, 14(4–5):505–525, 2001. DOI: 10.1016/S0893-6080(01)00020-X 20

R. Caruana. *Multitask Learning*. PhD thesis, Carnegie Mellon University, 1997. 87

R. Caruana and A. Niculescu-Mizil. An empirical comparison of supervised learning algorithms. In *Proc. of ICML*, 2006. DOI: 10.1145/1143844.1143865 2

X. Chen, L. Gu, S. Z. Li, and H.-J. Zhang. Learning representative local features for face detection. In *Proc. of CVPR*, 2001. DOI: 10.1090/S0002-9939-00-05692-6 8, 12

Y. Chen and Y. Rui. Real-time speaker tracking using particle filter sensor fusion. *Proceedings of the IEEE*, 92(3):485–494, 2004. DOI: 10.1109/JPROC.2003.823146 96

M. Collins, R. Schapire, and Y. Singer. Logistic regression, adaboost and bregman distances. *Machine Learning*, 48(1–3):253–285, 2002. DOI: 10.1023/A:1013912006537 15, 32, 78

R. Collins, A. Lipton, T. Kanade, H. Fujiyoshi, D. Duggins, Y. Tsin, D. Tolliver, N. Enomoto, and O. Hasegawa. A system for video surveillance and monitoring. Technical report, Robotics Institute, Carnegie Mellon University, 2000. 51

R. T. Collins and Y. Liu. Online selection of discriminative tracking features. In *ICCV*, 2003. DOI: 10.1109/TPAMI.2005.205 69, 76

D. Comaniciu, V. Ramesh, and P. Meer. Kernel-based object tracking. *IEEE Trans. on PAMI*, 25(5):564–577, 2003. DOI: 10.1109/TPAMI.2003.1195991 81

C. O. Conaire, N. E. O'Connor, and A. F. Smeaton. Detector adaptation by maximising agreement between independent data sources. In *Proc. of CVPR*, 2007. DOI: 10.1109/CVPR.2007.383448 77

N. Cristianini and J. Shawe-Taylor. *An Introduction to Support Vector Machines and other Kernel-Based Learning Methods*. Cambridge University Press, 2000. 23

F. Crow. Summed-area tables for texture mapping. In *Proc. of SIGGRAPH*, volume 18, pages 207–212, 1984. DOI: 10.1145/964965.808600 2

R. Cutler and L. Davis. Look who's talking: speaker detection using video and audio correlation. In *Proc. of IEEE ICME*, 2000. 95

R. Cutler, Y. Rui, A. Gupta, J. Cadiz, I. Tashev, L. He, A. Colburn, Z. Zhang, Z. Liu, and S. Silverbert. Distributed meetings: a meeting capture and broadcasting system. In *Proc. ACM Conf. on Multimedia*, 2002. DOI: 10.1145/641007.641112 94, 96, 107

N. Dalal and B. Triggs. Histogram of oriented gradients for human detection. In *Proc. of CVPR*, 2005. DOI: 10.1109/CVPR.2005.177 8, 12

T. G. Dietterich, R. H. Lathrop, and T. Lozano-Pérez. Solving the multiple instance problem with axis-parallel rectangles. *Artificial Intelligence*, 89(1–2):31–71, 1997. DOI: 10.1016/S0004-3702(96)00034-3 45

R. O. Duda, P. E. Hart, and D. G. Stork. *Pattern Classification*. John Wiley & Sons Inc., 2 edition, 2001. 17

S. Dupont and J. Luettin. Audio-visual speech modeling for continuous speech recognition. *IEEE Trans. on Multimedia*, 2(3):141–151, 2000. DOI: 10.1109/6046.865479 95

M. Enzweiler and D. M. Gavrila. Monocular pedestrian detection: Survey and experiments. *IEEE Trans. on PAMI*, 31(12):2179–2195, 2009. DOI: 10.1109/TPAMI.2008.260 12

R. Féraud, O. J. Bernier, J.-E. Viallet, and M. Collobert. A fast and accurate face detector based on neural networks. *IEEE Trans. on PAMI*, 23(1):42–53, 2001. DOI: 10.1109/34.899945 23, 24

J. Fisher III, T. Darrell, W. Freeman, and P. Viola. Learning joint statistical models for audio-visual fusion and segregation. In *NIPS*, pages 772–778, 2000. 95

Y. Freund and R. E. Schapire. A decision-theoretic generalization of on-line learning and an application to boosting. In *European Conf. on Computational Learning Theory*, 1994. DOI: 10.1006/jcss.1997.1504 4, 5, 15

Y. Freund and R. E. Schapire. A decision-theoretic generalization of on-line learning and an application to boosting. *Journal of Computer and System Sciences*, 55(1):119–139, 1997. DOI: 10.1006/jcss.1997.1504 4

J. Friedman, T. Hastie, and R. Tibshirani. Additive logistic regression: a statistical view of boosting. Technical report, Dept. of Statistics, Stanford University, 1998. DOI: 10.1214/aos/1016218223 3, 4, 17, 31, 32, 57, 75, 111

B. Fröba and A. Ernst. Fast frontal-view face detection using a multi-path decision tree. In *Proc. of Audio- and Video-based Biometric Person Authentication*, 2003. DOI: 10.1007/3-540-44887-X_107 16, 20, 21

B. Fröba and A. Ernst. Face detection with the modified census transform. In *IEEE Intl. Conf. on Automatic Face and Gesture Recognition*, 2004. DOI: 10.1109/AFGR.2004.1301514 8, 11

C. Galleguillos and S. Belongie. Context based object categorization: A critical survey. *Computer Vision and Image Understanding (CVIU)*, 114:712–722, 2010. DOI: 10.1016/j.cviu.2010.02.004 112

W. Gao, H. Ai, and S. Lao. Adaptive contour features in oriented granular space for human detection and segmentation. In *Proc. of CVPR*, 2009. DOI: 10.1109/CVPRW.2009.5206762 14

C. Garcia and M. Delakis. Convolutional face finder: A neural architecture for fast and robust face detection. *IEEE Trans. on PAMI*, 26(11):1408–1423, 2004. DOI: 10.1109/TPAMI.2004.97 23, 24

P. E. Gill, W. Murray, and M. H. Wright. *Practical Optimization*. Academic Press, 1981. 72

G. Iyengar and C. Neti. Speaker change detection using joint audio-visual statistics. In *The Int. RIAO Conference*, 2000. 95

S. Goodridge. *Multimedia sensor fusion for intelligent camera control and human computer interaction*. PhD thesis, Department of Electrical Engineering, North Carolina Start University, 1997. 99

H. Grabner and H. Bischof. On-line boosting and vision. In *Proc. of CVPR*, 2006. DOI: 10.1109/CVPR.2006.215 12, 14, 69, 76

G. Guo, H. Zhang, and S. Z. Li. Pairwise face recognition. In *Proc. of ICCV*, 2001. DOI: 10.1090/S0002-9939-00-05807-X 85, 87

T. Gustafsson, B. Rao, and M. Trivedi. Source localization in reverberant environments: performance bounds and ml estimation. In *Proc. of ICASSP*, 2001. DOI: 10.1109/ACSSC.2001.987753 97

F. Han, Y. Shan, H. S. Sawhney, and R. Kumar. Discovering class specific composite features through discriminative sampling with Swendsen-Wang cut. In *Proc. of CVPR*, 2008. DOI: 10.1109/CVPR.2008.4587376 14

D. Heckerman. A tractable inference algorithm for diagnosing multiple diseases. In *Proc. of UAI*, 1989. 45, 46, 47, 56

B. Heisele, T. Poggio, and M. Pontil. Face detection in still gray images. Technical report, Center for Biological and Computational Learning, MIT, A.I. Memo 1687, 2000. 24

B. Heisele, T. Serre, S. Prentice, and T. Poggio. Hierarchical classification and feature reduction for fast face detection with support vector machines. *Pattern Recognition*, 36:2007–2017, 2003. DOI: 10.1016/S0031-3203(03)00062-1 23, 24

B. Heisele, T. Serre, and T. Poggio. A component-based framework for face detection and identification. *International Journal of Computer Vision*, 74(2):167–181, 2007. DOI: 10.1007/s11263-006-0006-z 23, 25

J. Hershey and J. Movellan. Audio vision: using audio-visual synchrony to locate sounds. In *Advances in Neural Information Processing Systems*, 2000. 95

E. Hjelmas and B. K. Low. Face detection: A survey. *Computer Vision and Image Understanding*, 83: 236–274, 2001. DOI: 10.1006/cviu.2001.0921 1

K. Hotta. View independent face detection based on combination of local and global kernels. In *International Conference on Computer Vision Systems*, 2007. 23, 24

W. Hsu, S.-F. Chang, C.-W. Huang, L. Kennedy, C.-Y. Lin, and G. Iyengar. Discovery and fusion of salient multi-modal features towards news story segmentation. In *SPIE Electronic Imaging*, 2004. 95

C. Huang, H. Ai, Y. Li, and S. Lao. Vector boosting for rotation invariant multi-view face detection. In *Proc. of ICCV*, 2005. DOI: 10.1090/S0002-9939-04-07604-X 11, 16, 21, 55

C. Huang, H. Ai, Y. Li, and S. Lao. Learning sparse features in granular space for multi-view face detection. In *Intl. Conf. on Automatic Face and Gesture Recognition*, 2006. DOI: 10.1109/FGR.2006.70 8, 13, 19, 66

C. Huang, H. Ai, Y. Li, and S. Lao. High-performance rotation invariant multiview face detection. *IEEE Trans. on PAMI*, 29(4):671–686, 2007a. DOI: 10.1109/TPAMI.2007.1011 13, 21

C. Huang, H. Ai, T. Yamashita, S. Lao, and M. Kawade. Incremental learning of boosted face detector. In *Proc. of ICCV*, 2007b. DOI: 10.1109/ICCV.2007.4408850 77

G. B. Huang, R. Ramesh, T. Berg, and E. Learned-Miller. Labeled faces in the wild: A database for studying face recognition in unconstrained environments. Technical report, University of Massachusetts, Amherst, Technical Report 07-49, 2007c. 90, 93

G. B. Huang, M. Narayana, and E. Learned-Miller. Towards unconstrained face recognition. In *Proc. of IEEE Computer Society Workshop on Perceptual Organization in Computer Vision IEEE CVPR*, 2008. DOI: 10.1109/AFGR.1996.557268 85

V. Jain and E. Learned-Miller. FDDB: A benchmark for face detection in unconstrained settings. Technical report, University of Massachusetts, Amherst, 2010. 111

J.-S. Jang and J.-H. Kim. Fast and robust face detection using evolutionary pruning. *IEEE Trans. on Evolutionary Computation*, 12(5):562—571, 2008. DOI: 10.1109/TEVC.2007.910140 15, 16

O. Javed, S. Ali, and M. Shah. Online detection and classification of moving objects using progressively improving detectors. In *CVPR*, 2005. DOI: 10.1109/CVPR.2005.259 70, 77

O. Jesorsky, K. Kirchberg, and R. Frischholz. Robust face detection using the hausdorff distance. *Audio and Video based Person Authentication - AVBPA 2001*, pages 90–95, 2001. DOI: 10.1007/3-540-45344-X_14 60

H. Jin, Q. Liu, H. Lu, and X. Tong. Face detection using improved lbp under bayesian framework. In *Third Intl. Conf. on Image and Grahics (ICIG)*, 2004. 8, 11

M. Jones and P. Viola. Fast multi-view face detection. Technical report, Mitsubishi Electric Research Laboratories, TR2003-96, 2003. 8, 9, 10, 16, 20, 21, 24, 25, 54, 55, 67

M. Jones, P. Viola, and D. Snow. Detecting pedestrians using patterns of motion and appearance. Technical report, Mitsubishi Electric Research Laboratories, TR2003-90, 2003. 8, 9, 10, 51

B. Kapralos, M. Jenkin, and E. Milios. Audio-visual localization of multiple speakers in a video teleconferencing setting. Technical report, York University, Canada, 2002. 94, 96

J. D. Keeler, D. E. Rumelhart, and W.-K. Leow. Integrated segmentation and recognition of handprinted numerals. In *Proc. of NIPS*, 1990. 45, 46, 48

D. Keren, M. Osadchy, and C. Gotsman. Antifaces: A novel fast method for image detection. *IEEE Trans. on PAMI*, 23(7):747–761, 2001. DOI: 10.1109/34.935848 22, 23

T.-K. Kim and R. Cipolla. MCBoost: Multiple classifier boosting for perceptual co-clustering of images and visual features. In *Proc. of NIPS*, 2008. 16, 22, 26, 56

H. Kruppa, M. C. Santana, and B. Schiele. Fast and robust face finding via local context. In *Joint IEEE International Workshop on Visual Surveillance and Performance Evaluation of Tracking and Surveillance (VS-PETS)*, 2003. 112

I. Laptev. Improvements of object detection using boosted histograms. In *British Machine Vision Conference*, 2006. 12

B. Leibe, E. Seemann, and B. Schiele. Pedestrian detection in crowded scenes. In *Proc. of CVPR*, 2005. DOI: 10.1109/CVPR.2005.272 22

K. Levi and Y. Weiss. Learning object detection from a small number of examples: The importance of good features. In *Proc. of CVPR*, 2004. DOI: 10.1109/CVPR.2004.1315144 8, 12

A. Levin, P. Viola, and Y. Freund. Unsupervised improvement of visual detectors using co-training. In *ICCV*, 2003. 70, 77

S. Li, L. Zhu, Z. Zhang, A. Blake, H. Zhang, and H. Shum. Statistical learning of multi-view face detection. In *Proc. of ECCV*, 2002. DOI: 10.1090/S0002-9939-02-06510-3 4, 8, 9, 10, 15, 16, 20, 21, 30, 42, 54, 55, 67

S. Z. Li and Z. Zhang. Floatboost learning and statistical face detection. *IEEE Trans. on PAMI*, 26 (9):1112–1123, 2004. DOI: 10.1109/TPAMI.2004.68 9, 10, 15

Y. Li, S. Gong, and H. Liddell. Support vector regression and classification based multi-view face detection and recognition. In *International Conference on Automatic Face and Gesture Recognition*, 2000. DOI: 10.1109/AFGR.2000.840650 23, 24

Y. Li, H. Ai, T. Yamashita, S. Lao, and M. Kamade. Tracking in low frame rate video: a cascade particle filter with discriminative observers of different life spans. In *CVPR*, 2007. DOI: 10.1109/TPAMI.2008.73 69, 76

Y. Li, H. Ai, T. Yamashita, S. Lao, and M. Kamade. Tracking in low frame rate video: a cascade particle filter with discriminative observers of different life spans. *IEEE Trans. on PAMI*, 30(10): 1728–1740, 2008. DOI: 10.1109/TPAMI.2008.73 76

R. Lienhart and J. Maydt. An extended set of Haar-like features for rapid object detection. In *Proc. of ICIP*, 2002. DOI: 10.1109/ICIP.2002.1038171 8, 29

R. Lienhart, A. Kuranov, and V. Pisarevsky. Empirical analysis of detection cascades of boosted classifiers for rapid object detection. Technical report, Microprocessor Research Lab, Intel Labs, 2002. DOI: 10.1007/978-3-540-45243-0_39 15, 16, 18

Y.-Y. Lin and T.-L. Liu. Robust face detection with multi-class boosting. In *Proc. of CVPR*, 2005. DOI: 10.1090/S0002-9939-05-08264-X 16, 21, 58

C. Liu. A bayesian discriminating features method for face detection. *IEEE Trans. on PAMI*, 25(6): 725–740, 2003. DOI: 10.1109/TPAMI.2003.1201822 23

C. Liu and H.-Y. Shum. Kullback-Leibler boosting. In *Proc. Of CVPR*, 2003. DOI: 10.1090/S0002-9939-03-06881-3 8, 11, 12, 23

X. Liu and T. Yu. Gradient feature selection for online boosting. In *Proc. of ICCV*, 2007. DOI: 10.1090/S0002-9939-07-08711-4 14, 69, 76

J. Lu, K. N. Plataniotis, A. N. Venetsanopoulos, and S. Z. Li. Ensemble-based discriminant learning with boosting for face recognition. *IEEE Trans. on Neural Networks*, 17:166–178, 2006. DOI: 10.1109/TNN.2005.860853 85

H. Luo. Optimization design of cascaded classifiers. In *Proc. of CVPR*, 2005. DOI: 10.1090/S0002-9939-05-07941-4 16, 18, 30

O. Maron and T. Lozano-Perez. A framework for multiple-instance learning. In *Proc. of NIPS*, 1998. 46

H. Masnadi-Shirazi and N. Vasconcelos. Asymmetric boosting. In *Proc. of ICML*, 2007a. DOI: 10.1145/1273496.1273573 16, 17, 34

H. Masnadi-Shirazi and N. Vasconcelos. High detection-rate cascades for real-time object detection. In *Proc. of ICCV*, 2007b. DOI: 10.1109/ICCV.2007.4408860 17

L. Mason, J. Baxter, P. Bartlett, and M. Frean. Boosting algorithms as gradient descent. In *Proc. of NIPS*, 2000. 17, 45, 46, 47, 55, 56, 75, 76, 89, 111

B. McCane and K. Novins. On training cascade face detectors. In *Image and Vision Computing*, 2003. 16, 19

R. Meir and G. Rätsch. An introduction to boosting and leveraging. *S. Mendelson and A. J. Smola Ed., Advanced Lectures on Machine Learning, Springer-Verlag Berlin Heidelberg*, pages 118–183, 2003. DOI: 10.1007/3-540-36434-X 3

J. Meynet, V. Popovici, and J.-P. Thiran. Face detection with boosted gaussian features. *Pattern Recognition*, 40(8):2283–2291, 2007. DOI: 10.1016/j.patcog.2007.02.001 8, 12

K. Mikolajczyk, C. Schmid, and A. Zisserman. Human detection based on a probabilistic assembly of robust part detectors. In *Proc. of ECCV*, 2004. DOI: 10.1007/978-3-540-24670-1_6 23, 25

T. P. Minka. Algorithms for maximum-likelihood logistic regression. Technical report, Dept. of Statistics, Carnegie Mellon University, 2001. 72, 77

T. Mita, T. Kaneko, and O. Hori. Joint Haar-like features for face detection. In *Proc. of ICCV*, 2005. DOI: 10.1109/ICCV.2005.129 4, 10, 11, 15, 16

T. Mita, T. Kaneko, B. Stenger, and O. Hori. Discriminative feature co-occurrence selection for object detection. *IEEE Trans. on PAMI*, 30(7):1257–1269, 2008. DOI: 10.1109/TPAMI.2007.70767 8, 10

B. Moghaddam, T. Jebara, and A. Pentland. Bayesian face recognition. *IEEE Trans. on PAMI*, 33: 1771–1782, 2000. DOI: 10.1016/S0031-3203(99)00179-X 87

A. Mohan, C. Papageorgiou, and T. Poggio. Example-based object detection in images by components. *IEEE Trans. on PAMI*, 23(4):349—-361, 2001. DOI: 10.1109/34.917571 25

V. Nair and J. J. Clark. An unsupervised, online learning framework for moving object detection. In *CVPR*, 2004. DOI: 10.1109/CVPR.2004.1315181 70, 73, 77

M. Naphade, A. Garg, and T. Huang. Duration dependent input output Markov models for audio-visual event detection. In *Proc. of IEEE ICME*, 2001. DOI: 10.1109/ICME.2001.1237704 95

M. C. Nechyba, L. Brandy, and H. Schneiderman. Pittpatt face detection and tracking for the CLEAR 2007 evaluation. In *Classification of Events, Activities and Relations Evaluation and Workshop*, 2007. DOI: 10.1007/978-3-540-68585-2_10 25

K. Nickel, T. Gehrig, R. Stiefelhagen, and J. McDonough. A joint particle filter for audio-visual speaker tracking. In *ICMI*, 2005. DOI: 10.1145/1088463.1088477 96

H. Nock, G. Iyengar, and C. Neti. Speaker localisation using audio-visual synchrony: an empirical study. In *Proc. of CIVR*, 2003. DOI: 10.1007/3-540-45113-7_48 95

S. J. Nowlan and J. C. Platt. A convolutional neural network hand tracker. In *Proc. of NIPS*, volume 7, 1995. 18, 46, 111

T. Ojala, M. Pietikäinen, and T. Mäenpää. Multiresolution gray-scale and rotation invariant texture classification with local binary patterns. *IEEE Trans. on PAMI*, 24:971–987, 2002. DOI: 10.1109/TPAMI.2002.1017623 11, 85, 86

A. Opelt, A. Pinz, and A. Zisserman. A boundary-fragment-model for object detection. In *Proc. of CVPR*, 2006. DOI: 10.1007/11744047_44 8, 14

M. Osadchy, M. L. Miller, and Y. L. Cun. Synergistic face detection and pose estimation with energy-based models. In *Proc. of NIPS*, 2004. DOI: 10.1007/11957959_10 23, 24

E. Osuna, R. Freund, and F. Girosi. Training support vector machines: An application to face detection. In *Proc. of CVPR*, 1997. DOI: 10.1109/CVPR.1997.609310 24

N. C. Oza. *Online Ensemble Learning*. PhD thesis, University of Calfornia, Berkeley, 2002. 77

V. Pavlović, A. Garg, J. Rehg, and T. Huang. Multimodal speaker detection using error feedback dynamic Bayesian networks. In *Proc. of IEEE CVPR*, 2001. DOI: 10.1090/S0002-9939-01-05784-7 95

M.-T. Pham and T.-J. Cham. Detection caching for faster object detection. In *Proc. of CVPR*, 2005. 16, 20

M.-T. Pham and T.-J. Cham. Online learning asymmetric boosted classifiers for object detection. In *Proc. of CVPR*, 2007a. 16, 17, 76

M.-T. Pham and T.-J. Cham. Fast training and selection of haar features during statistics in boosting-based face detection. In *Proc. of ICCV*, 2007b. 16, 19

P. J. Phillips, H. Moon, P. J. Rauss, and S. Rizvi. The FERET evaluation methodology for face recognition algorithms. *IEEE Trans. on PAMI*, 22(10):1090–1104, 2000. DOI: 10.1109/34.879790 60

F. Porikli. Integral histogram: A fastway to extract histograms in cartesian spaces. In *Proc. of CVPR*, 2005. 19

P. Pudil, J. Novovicova, and J. Kittler. Floating search methods in feature selection. *Pattern Recognition Letters*, 15(11):1119–1125, 1994. DOI: 10.1016/0167-8655(94)90127-9 15

M. Rätsch, S. Romdhani, and T. Vetter. Efficient face detection by a cascaded support vector machine using haar-like features. In *Pattern Recognition Symposium*, 2004. 23, 24

S. Romdhani, P. Torr, B. Schölkopf, and A. Blake. Computationally efficient face detection. In *Proc. of ICCV*, 2001. DOI: 10.1109/ICCV.2001.937694 23, 24

D. Roth, M.-H. Yang, and N. Ahuja. A SNoW-based face detector. In *Proc. of NIPS*, 2000. 24

P. M. Roth, H. Grabner, D. Skočaj, H. Bischof, and A. Leonardis. Online conservative learning for person detection. In *Proc. 2nd Joint IEEE International Workshop on VS-PETS*, 2005. 73, 77

H. Rowley, S. Baluja, and T. Kanade. Neural network-based face detection. *IEEE Trans. on PAMI*, 20:23–38, 1998. DOI: 10.1109/34.655647 41, 62

H. A. Rowley, S. Baluja, and T. Kanade. Neural network-based face detection. In *Proc. of CVPR*, 1996. DOI: 10.1109/CVPR.1996.517075 24

H. A. Rowley, S. Baluja, and T. Kanade. Rotation invariant neural network-based face detection. Technical report, School of Coomputer Science, Carnegie Mellow Univ., CMU-CS-97-201, 1997. DOI: 10.1109/CVPR.1998.698721 24

Y. Rui, D. Florencio, W. Lam, and J. Su. Sound source localization for circular arrays of directional microphones. In *Proc. of IEEE ICASSP*, 2005. DOI: 10.1109/ICASSP.2005.1415654 94, 101

P. Sabzmeydani and G. Mori. Detecting pedestrians by learning shapelet features. In *Proc. of CVPR*, 2007. 8, 14

R. E. Schapire and Y. Singer. Improved boosting algorithms using confidence-rated predictions. *Machine Learning*, 37:297–336, 1999. DOI: 10.1023/A:1007614523901 4, 5, 55, 57, 111

H. Schneiderman. Learning a restricted bayesian network for object detection. In *Proc. of CVPR*, 2004a. DOI: 10.1109/CVPR.2004.1315224 23, 25

H. Schneiderman. Feature-centric evaluation for efficient cascaded object detection. In *Proc. of CVPR*, 2004b. DOI: 10.1109/CVPR.2004.1315141 16, 19, 25

H. Schneiderman and T. Kanade. A statistical model for 3d object detection applied to faces and cars. In *Proc. of CVPR*, 2000. 62

H. Schneiderman and T. Kanade. Object detection using the statistics of parts. *International Journal of Computer Vision*, 56(3):151–177, 2004. DOI: 10.1023/B:VISI.0000011202.85607.00 23, 25

E. Seemann, B. Leibe, and B. Schiele. Multi-aspect detection of articulated objects. In *Proc. of CVPR*, 2006. DOI: 10.1109/CVPR.2006.193 16, 22, 54

Y. Shan, F. Han, H. S. Sawhney, and R. Kumar. Learning exemplar-based categorization for the detection of multi-view multi-pose objects. In *Proc. of CVPR*, 2006. DOI: 10.1109/CVPR.2006.168 16, 22

J. Shotton, A. Blake, and R. Cipolla. Contour-based learning for object detection. In *Proc. of ICCV*, 2005. DOI: 10.1109/ICCV.2005.63 8, 14

T. Sim, S. Baker, and M. Bsat. The CMU pose, illumination, and expression database. *IEEE Trans. on PAMI*, 25(12):1615–1618, 2003. DOI: 10.1109/TPAMI.2003.1251154 60, 61

J. Sochman and J. Matas. Waldboost - learning for time constrained sequential detection. In *Proc. of CVPR*, 2005. DOI: 10.1109/CVPR.2005.373 16, 18, 29, 42

Z. Stone, T. Zickler, and T. Darrell. Autotagging facebook: Social network context improves photo annotation. In *Proceedings of CVPR Workshop on Internet Vision*, 2008. 85

F. Suard, A. Rakotomamonjy, A. Bensrhair, and A. Broggi. Pedestrian detection using infrared images and histograms of oriented gradients. In *IEEE Intelligent Vehicles Symposium*, 2006. DOI: 10.1109/IVS.2006.1689629 12

K. Sung and T. Poggio. Example-based learning for view-based face detection. *IEEE Trans. on PAMI*, 20:39–51, 1998. DOI: 10.1109/34.655648 41

A. Torralba, K. P. Murphy, and W. T. Freeman. Sharing features: Efficient boosting procedures for multiclass object detection. In *Proc. of CVPR*, 2004. DOI: 10.1109/CVPR.2004.1315241 21, 22, 26

A. Torralba, K. P. Murphy, and W. T. Freeman. Sharing visual features for multiclass and multiview object detection. *IEEE Trans. on PAMI*, 29(5):854–869, 2007. DOI: 10.1109/TPAMI.2007.1055 57

Z. Tu. Probabilistic boosting-tree: Learning discriminative models for classification, recognition, and clustering. In *Proc. of ICCV*, 2005. DOI: 10.1109/ICCV.2005.194 16, 22

O. Tuzel, F. Porikli, and P. Meer. Region covariance: A fast descriptor for detection and classification. In *Proc. of ECCV*, 2006. DOI: 10.1007/11744047_45 8, 12

J. Vermaak, M. Gangnet, A. Black, and P. Pérez. Sequential Monte Carlo fusion of sound and vision for speaker tracking. In *Proc. of IEEE ICCV*, 2001. DOI: 10.1109/ICCV.2001.937600 95, 96

R. Verschae, J. Ruiz-del Solar, and M. Correa. Face recognition in unconstrained environments: A comparative study. In *Proc. of ECCV Workshop on Faces in Real-Life Images*, 2008. 85

P. Viola and M. Jones. Rapid object detection using a boosted cascade of simple features. In *Proc. of CVPR*, 2001. DOI: 10.1109/CVPR.2001.990517 xi, 2, 6, 7, 8, 11, 15, 16, 17, 22, 23, 29, 30, 34, 41, 42, 66, 67, 105, 111

P. Viola and M. Jones. Fast and robust classification using asymmetric AdaBoost and a detector cascade. In *Proc. of NIPS*, 2002. 15, 16, 34

P. Viola, J. C. Platt, and C. Zhang. Multiple instance boosting for object detection. In *Proc. of NIPS*, volume 18, 2005. xi, 22, 26

A. Wald. *Sequential Analysis*. Dover, 1947. 18

H. Wang and P. Chu. Voice source localization for automatic camera pointing system in video-conferencing. In *Proc. of IEEE ICASSP*, 1997. DOI: 10.1090/S0002-9939-97-03935-X 94, 96, 101

P. Wang and Q. Ji. Multi-view face detection under complex scene based on combined svms. In *Proc. of ICPR*, 2004. DOI: 10.1090/S0002-9939-04-07414-3 23, 24

P. Wang and Q. Ji. Learning discriminant features for multi-view face and eye detection. In *Proc. of CVPR*, 2005. DOI: 10.1090/S0002-9939-04-07669-5 8, 12

X. Wang, T. X. Han, and S. Yan. An HOG-LBP human detector with partial occlusion handling. In *Proc. of ICCV*, 2009a. 8, 12

X. Wang, C. Zhang, and Z. Zhang. Boosted multi-task learning for face verification with applications to web image and video search. In *Proc. of CVPR*, 2009b. DOI: 10.1109/CVPRW.2009.5206736 xi, 26

C. A. Waring and X. Liu. Face detection using spectral histograms and SVMs. *IEEE Trans. on Systems, Man, and Cybernetics – Part B: Cybernetics*, 35(3):467—476, 2005. DOI: 10.1109/TSMCB.2005.846655 8, 12

A. R. Webb. *Statistical Pattern Recognition*. Oxford University Press, 1 edition, 1999. 19

B. Wu and R. Nevatia. Detection of multiple, partially occluded humans in a single image by bayesian combination of edgelet part detectors. In *Proc. of ICCV*, 2005. DOI: 10.1090/S0002-9939-04-07528-8 8, 14

B. Wu and R. Nevatia. Cluster boosted tree classifier for multi-view, multi-pose object detection. In *Proc. of ICCV*, 2007a. DOI: 10.1109/ICCV.2007.4409006 16, 22, 54, 55, 60

B. Wu and R. Nevatia. Simultaneous object detection and segmentation by boosting local shape feature based classifier. In *Proc. of CVPR*, 2007b. DOI: 10.1109/CVPR.2007.383042 14

B. Wu, H. Ai, C. Huang, and S. Lao. Fast rotation invariant multi-view face detection based on real adaboost. In *Proc. of IEEE Automatic Face and Gesture Recognition*, 2004a. DOI: 10.1109/AFGR.2004.1301512 4, 11, 15, 16, 20, 21, 33, 35, 42, 54, 55, 65, 66, 67

J. Wu, J. M. Rehg, and M. D. Mullin. Learning a rare event detection cascade by direct feature selection. In *Proc. of NIPS*, volume 16, 2004b. 16, 19

J. Wu, S. C. Brubaker, M. D. Mullin, and J. M. Rehg. Fast asymmetric learning for cascade face detection. Technical report, Georgia Institute of Technology, GIT-GVU-05-27, 2005. 16, 17, 34

J. Wu, S. C. Brubaker, M. D. Mullin, and J. M. Rehg. Fast asymmetric learning for cascade face detection. *IEEE Trans. on PAMI*, 30(3):369–382, 2008. DOI: 10.1109/TPAMI.2007.1181 17

R. Xiao, L. Zhu, and H. Zhang. Boosting chain learning for object detection. In *Proc. of ICCV*, 2003. DOI: 10.1090/S0002-9939-02-06891-0 15, 16, 30, 42, 66, 67

R. Xiao, H. Zhu, H. Sun, and X. Tang. Dynamic cascades for face detection. In *Proc. of ICCV*, 2007. DOI: 10.1090/S0002-9939-06-08537-6 11, 16, 18, 30, 33

J. Yagnik and A. Islam. Learning people annotation from the web via consistency learning. In *Proc. of the International Workshop on Multimedia Information Retrieval*, 2007. DOI: 10.1145/1290082.1290121 85

J. Yan, S. Li, S. Zhu, and H. Zhang. Ensemble svm regression based multi-view face detection system. Technical report, Microsoft Research, MSR-TR-2001-09, 2001. 23, 24

S. Yan, S. Shan, X. Chen, and W. Gao. Locally assembled binary (LAB) feature with feature-centric cascade for fast and accurate face detection. In *Proc. of CVPR*, 2008. DOI: 10.1109/CVPR.2008.4587802 8, 11, 16, 20

M. Yang, Y. Wu, and G. Hua. Context-aware visual tracking. *IEEE Trans. on PAMI*, 31(7):1195–1209, 2009. DOI: 10.1109/TPAMI.2008.146 112

M.-H. Yang, D. J. Kriegman, and N. Ahuja. Detecting faces in images: A survey. *IEEE Trans. on PAMI*, 24(1):34–58, 2002. DOI: 10.1109/34.982883 1, 2, 22, 23, 25

P. Yang, S. Shan, W. Gao, S. Z. Li, and D. Zhang. Face recognition using adaboosted gabor features. In *Proc. of Intl. Conf. on Face and Gesture Recognition*, 2004. DOI: 10.1090/S0002-9939-03-07088-6 85, 87

B. Yoshimi and G. Pingali. A multimodal speaker detection and tracking system for teleconferencing. In *Proc. ACM Conf. on Multimedia*, 2002. DOI: 10.1145/641007.641100 94, 96

J. Yuan, J. Luo, and Y. Wu. Mining compositional features for boosting. In *Proc. of CVPR*, 2008. DOI: 10.1109/CVPR.2008.4587347 13

C. Zhang and P. Viola. Multiple-instance pruning for learning efficient cascade detectors. In *Proc. of NIPS*, 2007. DOI: 10.1090/S0002-9939-06-08821-6 xi, 16, 26, 62, 67, 90

C. Zhang and Z. Zhang. Winner-take-all multiple category boosting for multi-view face detection. Technical report, Microsoft Research MSR-TR-2009-190, 2009. xi

C. Zhang, Z. Zhang, and D. Florêncio. Maximum likelihood sound source localization for multiple directional microphones. In *ICASSP*, 2007a. DOI: 10.1109/ICASSP.2007.366632 96, 98

C. Zhang, R. Hamid, and Z. Zhang. Taylor expansion based classifier adaptation: Appiication to person detection. In *Proc. of CVPR*, 2008a. DOI: 10.1109/CVPR.2008.4587801 xi, 26

C. Zhang, P. Yin, Y. Rui, R. Cutler, P. Viola, X. Sun, N. Pinto, and Z. Zhang. Boosting-based multimodal speaker detection for distributed meeting videos. *IEEE Trans. on Multimedia*, 10(8): 1541–1552, 2008b. DOI: 10.1109/TMM.2008.2007344 xi, 27

G. Zhang, X. Huang, S. Z. Li, Y. Wang, and X. Wu. Boosting local binary pattern (LBP)-based face recognition. In *Proc. Advances in Biometric Person Authentication*, 2004. 11, 85, 87, 90, 92

H. Zhang, W. Gao, X. Chen, and D. Zhao. Object detection using spatial histogram features. *Image and Vision Computing*, 24(4):327–341, 2006. DOI: 10.1016/j.imavis.2005.11.010 8, 12

L. Zhang, R. Chu, S. Xiang, S. Liao, and S. Z. Li. Face detection based on multi-block LBP representation. 2007b. DOI: 10.1007/978-3-540-74549-5_2 8, 11

Q. Zhu, S. Avidan, M.-C. Yeh, and K.-T. Cheng. Fast human detection using a cascade of histograms of oriented gradients. In *Proc. of CVPR*, 2006. DOI: 10.1090/S0002-9939-05-08118-9 12

X. Zhu. Semi-supervised learning literature survey. Technical report, 1530, Dept. of Computer Science, University of Wisconsin - Madison, 2007. 77

D. N. Zotkin, R. Duraiswami, and L. S. Davis. Logistic regression, adaboost and bregman distances. *EURASIP Journal on Applied Signal Processing*, 2002(11):1154–1164, 2002. DOI: 10.1155/S1110865702206058 95, 96

Authors' Biographies

CHA ZHANG

Cha Zhang is a Researcher in the Communication and Collaboration Systems Group at Microsoft Research (Redmond, WA). He received the B.S. and M.S. degrees from Tsinghua University, Beijing, China in 1998 and 2000, respectively, both in Electronic Engineering, and the Ph.D. degree in Electrical and Computer Engineering from Carnegie Mellon University, in 2004. His current research focuses on applying various machine learning and computer graphics/computer vision techniques to multimedia applications, in particular, multimedia teleconferencing. During his graduate studies at CMU, he worked on various multimedia related projects including sampling and compression of image-based rendering data, 3D model database retrieval and active learning for database annotation, peer-to-peer networking, etc. Dr. Zhang has published more than 40 technical papers and holds 10+ U.S. patents. He won the best paper award at ICME 2007, the top 10% award at MMSP 2009, and the best student paper award at ICME 2010. He co-authored a book titled Light Field Sampling, published by Morgan and Claypool in 2006.

Dr. Zhang is a Senior Member of IEEE. He was the Publicity Chair for International Packet Video Workshop in 2002, the Program Co-Chair for the first Immersive Telecommunication Conference (IMMERSCOM) in 2007, the Steering Committee Co-Chair and Publicity Chair for IMMERSCOM 2009, the Program Co-Chair for the ACM Workshop on Media Data Integration (in conjunction with ACM Multimedia 2009), and the Poster&Demo Chair for ICME 2011. He served as TPC members for many conferences including ACM Multimedia, CVPR, ICCV, ECCV, MMSP, ICME, ICPR, ICWL, etc. He served as an Associate Editor for Journal of Distance Education Technologies, IPSJ Transactions on Computer Vision and Applications, and ICST Transactions on Immersive Telecommunications. He was a guest editor for Advances in Multimedia, Special Issue on Multimedia Immersive Technologies and Networking.

ZHENGYOU ZHANG

Zhengyou Zhang received the B.S. degree in electronic engineering from the University of Zhejiang, Hangzhou, China, in 1985, the M.S. degree in computer science from the University of Nancy, Nancy, France, in 1987, and the Ph.D. degree in computer science and the Doctorate of Science (*Habilitation à diriger des recherches*) from the University of Paris XI, Paris, France, in 1990 and 1994, respectively.

He is a Principal Researcher with Microsoft Research, Redmond, WA, USA, and he manages the multimodal collaboration research team. Before joining Microsoft Research in March 1998, he was with INRIA (French National Institute for Research in Computer Science and Control), France, for 11 years and was a Senior Research Scientist from 1991. In 1996-1997, he spent a one-year sabbatical as an Invited Researcher with the Advanced Telecommunications Research Institute International (ATR), Kyoto, Japan. He has published over 200 papers in refereed international journals and conferences, and he has coauthored the following books: *3-D Dynamic Scene Analysis: A Stereo Based Approach* (Springer-Verlag, 1992); *Epipolar Geometry in Stereo, Motion and Object Recognition* (Kluwer, 1996); *Computer Vision* (Chinese Academy of Sciences, 1998, 2003, in Chinese); and *Face Geometry and Appearance Modeling* (Cambridge University Press, 2010, to appear). He has given a number of keynotes in international conferences.

Dr. Zhang is a Fellow of the *Institute of Electrical and Electronic Engineers* (IEEE), the Founding Editor-in-Chief of the *IEEE Transactions on Autonomous Mental Development*, an Associate Editor of the *International Journal of Computer Vision*, and an Associate Editor of *Machine Vision and Applications*. He served as Associate Editor of the *IEEE Transactions on Pattern Analysis and Machine Intelligence* from 2000 to 2004, an Associate Editor of the *IEEE Transactions on Multimedia* from 2004 to 2009, among others. He has been on the program committees for numerous international conferences in the areas of autonomous mental development, computer vision, signal processing, multimedia, and human-computer interaction. He is a Program Co-Chair of the *International Conference on Multimedia and Expo* (ICME), July 2010, a Program Co-Chair of the ACM *International Conference on Multimedia* (ACM MM), October 2010, and a Program Co-Chair of the ACM *International Conference on Multimodal Interfaces* (ICMI), November 2010.